The Information Retrieval Series, Vol. 26

Series Editor

W. Bruce Croft

Victor Lavrenko

A Generative Theory of Relevance

Victor Lavrenko
University of Edinburgh
School of Informatics
10 Crichton Street
Edinburgh
United Kingdom EH8 9AB
v.lavrenko@gmail.com

ISBN: 978-3-642-10042-0 e-ISBN: 978-3-540-89364-6

ACM Computing Classification (1998): H.3, G.3, F.2

Softcover reprint of the hardcover 1st edition 2009

Cover design: Künkel Lopka, Heidelberg

Printed on acid-free paper

9 8 7 6 5 4 3 2 1

springer.com

To Daria and Dar

Preface

The purpose of this book is to introduce a formal way of thinking about the notion of relevance in the field of Information Retrieval. Relevance is viewed as a generative process, and we hypothesize that both user queries and relevant documents represent random observations from that process. Based on this view, we develop a formal retrieval model that has direct applications to a wide range of search scenarios. The new model substantially outperforms strong baselines on the tasks of ad-hoc retrieval, cross-language retrieval, handwriting retrieval, automatic image annotation, video retrieval, structured search and topic detection and tracking. Empirical success of our approach is due to a new technique we propose for modeling exchangeable sequences of discrete random variables. The new technique represents an attractive counterpart to existing formulations, such as multinomial and simplicial mixtures: it is effective, easy to train, and makes no assumptions about the geometric structure of the data.

Edinburgh, Scotland
September 2008

Victor Lavrenko

Acknowledgments

This book would not have been possible without the help of my relatives, mentors and colleagues. I was fortunate to work closely with Bruce Croft, who helped stimulate many of the ideas presented in this book. I am also tremendously thankful to James Allan, who guided and advised me since the early days of my undergraduate studies. I would like to thank my fellow students and colleagues at the University of Massachusetts Amherst. Early discussions with Warren Greiff and Anton Leuski had a strong impact on my way of thinking about the field of information retrieval. I thank Andres Corrada-Emmanuel for his unorthodox point of view and willingness to challenge assumptions. Experimental results presented in this book are a product of very enjoyable collaborations with my colleagues at UMass. Cross-language experiments were carried out in collaboration with Martin Choquette and Bruce Croft. Multimedia research was a joint work with R. Manmatha, Toni Rath, Jiwoon Jeon and Shaolei Feng. Structured retrieval was a collaborative effort with James Allan and Xing Yi. Topic Detection and Tracking work was done with James Allan with the aid of Margaret Connell and Leah Larkey. I want to give special thanks to my wife for her understanding, patience and unending support.

Contents

List of Figures

List of Tables

1

Introduction

Information Retrieval (IR) is a field of science concerned with searching for useful information in large, loosely structured or unstructured collections. In the 1970s, when the field was in its infancy, the word *information* primarily meant *text*, specifically the kinds of text one might find in a library. Today information exists in many more forms, often quite different from books or scientific articles which were the focus of early retrieval systems. Web searching is perhaps the most famed application of information retrieval today, and on the web we may find relevant information in the form of text, but also as images, as audio files, as video segments, even as hyper-links between various pages. A modern search system must have the capability to find, organize and present to the user all of these very different manifestations of information, because all of them may have some relevance to the user's information need.

The notion of *relevance* plays an extremely important part in Information Retrieval, some went so far as to say that it defines the entire field and serves as a distinguishing feature from database theory and library science [120]. The concept of relevance, while seemingly intuitive, is nevertheless hard to define, and even harder to model in a formal fashion, as evidenced by over 160 publications attempting to deal with the issue. Relevance is also the subject of this book. We will not attempt to bring forth a new definition of relevance, nor will we provide arguments as to why one definition might be theoretically superior or more complete than another. Instead, we will take a widely accepted, albeit somewhat conservative definition, make several assumptions, and from them develop a new probabilistic model that explicitly captures that notion of relevance. In our development we will make a determined effort to avoid the heuristic steps that accompanied the development of popular retrieval models. To this end, a substantial portion of our narrative will be devoted to a formal discussion of event spaces and assumptions underlying our model.

1.1 Contributions

There are two major contributions in this work. The first contribution is a new way to look at topical relevance within the field of Information Retrieval. The second contribution is a new generative model for collections of discrete data.

1.1.1 A new model of relevance

Our first contribution will be of primary interest to a researcher in Information Retrieval, or to anyone concerned with modeling topical content. This book will propose a new way to look at topical relevance. The new approach will be summarized as a hypothesis (defined in section 3.1), and will lead to a new formal model of Information Retrieval. Some of the main advantages of the model are:

- It represents a natural complement of the two dominant models of retrieval: the classical probabilistic model and the language modeling approach.
- The model *explicitly* combines documents, queries and relevance in a single formalism. This has been somewhat of a challenge in previous formulations.
- It requires no heuristics to compute the probabilities associated with the relevant class.
- The retrieval model has relevance feedback and query expansion as integral parts.
- It captures a wide array of IR tasks. As far as we know, this is the first model capable of performing ad-hoc search, image and video annotation, handwriting retrieval, semi-structured search and Topic Detection all within the same formalism.
- It outperforms state-of-the-art benchmarks. A few examples: 10-25% improvement in ad-hoc search, 5-30% in topic detection, 15-60% in image annotation and retrieval.
- It allows queries to be issued in any form: keywords, passages, whole documents, sketches, portions of images, video frames – as long as we have enough data to estimate the parameters.
- It can recover missing portions of incomplete or degraded information representations and provides a particularly effective method of searching semi-structured databases which achieves a 30% improvement over a language-modeling baseline.

1.1.2 A new generative model

The second contribution of our work may be useful to a language-modeling practitioner, or to a reader with a more general interest in statistical modeling (not necessarily of language). In chapter 4 we will propose a new method for modeling exchangeable sequences of discrete random variables. The main distinguishing qualities of our formalism are:

- It can successfully handle rare events and does not ignore outliers in the training data.
- It does not make any structural assumptions, allowing the data to speak for itself.
- The model is easy to train, either generatively, or with a discriminative objective.
- The model is effective: 15-20% reduction in predictive perplexity over the baselines, plus successful applications to retrieval tasks mentioned in the previous section.

1.1.3 Minor contributions

In addition to the two major contributions, this book contains a number of smaller results. These results mostly represent new insights into existing models, or serve as explanations for curious experimental observations. Some of these minor contributions are:

- We will show that, contrary to popular belief, the classical probabilistic model need not be predicated on the assumption of *word independence*, or even on the weaker assumption of *linked dependence* suggested by [29]. Instead, it requires a much more plausible assumption, which we call *proportional interdependence.*
- We will try to explain why explicit models of word dependence have never resulted in consistent improvements in retrieval performance, either in the classical or the language-modeling framework.
- We will assert that documents and queries can be viewed as samples from the same underlying distribution, even though they may look very different.
- We will explain why relative entropy (KL divergence) has empirically outperformed probability ratio as a criterion for ranking documents in the language-modeling framework.
- We will conclude that a Probabilistic LSI model with k aspects is equivalent to a regular, document-level mixture model with $N >> k$ components.
- We will argue that Latent Dirichlet Allocation is not a word-level mixture, it is a regular document-level mixture restricted to the k-topic sub-simplex.
- We will demonstrate that a form of KL-divergence ranking may be interpreted as a statistical significance test for a hypothesis that a document and a query originated from the same relevant population.

1.2 Overview

This book will be structured in the following manner. **Chapter 2** will be devoted to the notion of relevance and to previously proposed models of relevance. We will start with a simple definition and then briefly outline several arguments challenging that definition. Sections 2.1.2 through 2.1.5 will contain several alternative views of relevance. Conceptual discussions will culminate in section 2.2, where we will describe Mizarro's attempt [92] to construct a unified definition of relevance. In section 2.3 we will discuss two main approaches for dealing with relevance: the classical probabilistic framework (section 2.3.2) and the language modeling framework (section 2.3.3). Our discussion will focus heavily on the probabilistic representations adopted by the two models, and on the modeling assumptions made in each case. Sections 2.3.2 and 2.3.3 contain novel arguments suggesting that modeling word dependence may be a futile endeavor, at least if one is concerned with topical relevance. We conclude chapter 2 by contrasting the two frameworks and pointing out that both are not completely adequate: the classical framework because of its reliance on heuristics, and the language modeling framework because of the explicit absence of relevance in the model. We will also briefly discuss the risk-minimization framework advocated by Zhai [157] as a possible work-around for the shortcomings of the language-modeling framework.

Chapter 3 contains the main theoretical ideas of our work. In this chapter we will introduce a new framework for viewing relevance and describe how this framework can be applied in a number of search scenarios. In section 3.1 we will provide an informal overview of our ideas. We will also define the core proposition of our work: the *generative relevance hypothesis.* Section 3.2 and on will contain a formal development of our framework. Section 3.3 will focus on representation of documents, queries and relevance, as well as on functional relations between these representations. Section 3.4 will discuss probability distributions associated with relevance, and how these distributions affect document and query observations. In section 3.6 we describe how our model can be used for ranking documents in response to a query. We will discuss two ranking criteria: one based on the probability ranking principle, the other rooted in statistical hypothesis testing. We will also discuss relations between these criteria in section 3.6.4 and provide an original argument for why probability-ratio scoring may lead to sub-optimal document ranking (section 3.6.3). Section 3.7 concludes the chapter and contrasts our model against both the classical probabilistic model and the language modeling framework.

Chapter 4 is dedicated to the statistical machinery underlying our model. The methods discussed in this chapter are very general and extend well beyond the domain of Information Retrieval. We consider a deceivingly simple problem of modeling exchangeable sequences of discrete random variables, commonly known as *bags of words.* In section 4.2 we review five existing approaches to this problem, starting with a simple unigram model and progressing to advanced models like Latent Semantic Indexing (pLSI) and Latent Dirichlet Allocation (LDA). In section 4.3 we demonstrate that all five models can be expressed in a simple common form that is a direct consequence of De-Finetti's representation theorem. We draw several unexpected conclusions, showing that pLSI is equivalent to a simple discrete mixture, and LDA is a restricted form of a simple continuous mixture. Section 4.4 will introduce a new family of generative models, predicated on kernel-based density allocation within the De-Finetti representation. We will discuss two types of kernels: Dirac delta functions and Dirichlet kernels. We will also give an extensive argument for why we believe the new family may be superior to existing formulations.

Chapter 5 takes the general ideas from chapters 3 and 4 and shows how they can be turned into operational search systems. We discuss application of our framework to eight very different retrieval scenarios:

1. Ad-hoc retrieval: similar to simple web search, without the hyperlink information.
2. Relevance feedback: simulates interaction with the user to improve searching accuracy.
3. Cross-language retrieval: uses English queries to find documents in Chinese.
4. Handwriting retrieval: keyword search of degraded historical manuscripts.
5. Image annotation: we automatically annotate images with related keywords.
6. Video retrieval: search through a video collection using a text query.
7. Structured search: find relevant records in a database with missing field values.
8. Topic Detection and Tracking: organize live news-feeds into event-based groups.

In each case we will start by describing how the scenario fits into our model of relevance. We will then provide a detailed empirical evaluation, showing the effectiveness of our model against state-of-the-art baselines.

Chapter 6 will conclude the book. We will discuss the implications of our experiments and make suggestions for further development of our framework.

2

Relevance

2.1 The many faces of relevance

The notion of relevance serves as the foundation for the field of Information Retrieval. After all, the purpose of retrieval systems is to retrieve relevant items in response to user requests. Naturally, most users have a fairly good idea of what relevance is – it is a representation of their *information need*, a reflection of *what they are searching for*. However, in order to build and test effective retrieval systems we must translate the intuitive notion of relevance into a strict formalism, and that turns out to be somewhat of a challenge.

2.1.1 A simple definition of relevance

One of the simplest, and also most widely used, definitions of relevance is that of a binary relation between a given information item (document D) and the user's request (query Q). We might assume that the document is represented by a set of key words, appropriately reflecting its contents. Similarly, the user's request Q is a set of key words that represent the user's interest. Given these representations, we may say that a relevance relation between D and Q holds if there is a substantial overlap in meaning between the keyword sets of D and Q. Under this view, relevance of D does not depend on any factors other than representations of D and Q. Specifically, it does not depend on the user who issued the request, the task that prompted the request, or on user's preferences and prior knowledge. Similarly, in this simple definition relevance does not depend on any other documents D' in the collection, whether or not they have been examined by the user, or even judged relevant or non-relevant. It also does not depend on any other requests Q' to which D was previously judged relevant or non-relevant.

When relevance is defined as above, it is often called *system-oriented* or *algorithmic* relevance. The definition has been challenged and deemed inadequate on numerous occasions. In his comprehensive review of various formulations of relevance Mizarro [91] cites 160 papers attempting to define various aspects of relevance. In this section we will briefly highlight a few of the most popular arguments about the nature of relevance. A reader yearning for a more complete exposition of the subject is invited to examine the work of Saracevic [120], Robertson [113], Harter [53] and Mizarro [91].

2.1.2 User-oriented views of relevance

Discussions of the proper definition of relevance began in 1959 when Vickery [142, 143] argued for a distinction between "relevance to a subject" and "user relevance". The former refers to a degree of semantic correspondence between the user's request and an item returned by the system in response to that request. The latter is a reflection of how much the user likes the retrieved item, taking into account his task and previously seen items. The distinction between the two can be explained if we imagine that our collection contains two near duplicates of the same item. If one of them is semantically relevant to the request, so will be the other. However, the user may not be satisfied with the second item, since it is completely redundant.

A somewhat different dichotomy of relevance was mentioned by Maron and Kuhns, who incidentally were the first to treat relevance probabilistically. In [85] they consider the distinction between the user's request and the underlying *information need*. The request is a surface representation of information need, it is observable and readily available to the system and to other users. The information need itself is an abstract concept that only the user himself is aware of. A document that appears relevant to the request may be completely irrelevant to the underlying information need. This happens frequently because of ambiguity inherent in human communication; the problem is further aggravated by the users' tendency to keep their requests short and devoid of little redundancies that would be so helpful to a search engine.

Belkin et al. [9, 10] make a further abstraction of a user's *information need*. They introduce a concept of ASK (*Anomalous State of Knowledge*), signifying the fact that the user himself may not be fully aware of what he is searching for. The concept of information need is completely abstract, it is not observable; the user has only a perception of that need, and that perception can change during the course of a searching session. The same idea is raised in a number of other publications. For example, Ingwersen [59], coins the acronyms ISK and USK, referring respectively to *Incomplete* and *Uncertain* States of Knowledge.

Foskett [46, 47] and later Lancaster [73] make an interesting argument that the nature of relevance to a large degree depends on the person who is making a judgment. In their definition, the term *relevance* refers to a "public" or "social" notion, where the judgment is made by an external expert or collective, and not by the user who posed the request in the first place. For the case when the judgment is made by the user himself, they coin the term *pertinence*. In light of previous discussion by Maron and Kuhns [85], relevance is a relation between a document and the request, while pertinence is a relation between a document and the underlying information need.

In opposition to most of the distinctions drawn above, Fairthorne [43] makes a sobering argument for a strict definition of relevance, involving only the words contained in the document and the query. Otherwise, he claims, for any given document and any request, no matter how distant, we could hypothesize a situation where that document would in fact be relevant to the request.

2.1.3 Logical views of relevance

A number of authors attempted to define relevance formally, through logical constructs on the semantics of the documents and requests. One of the first formal attempts is due to Cooper [25], who defines relevance in terms of entailment (as in theorem-proving). Suppose q represents a logical proposition corresponding to a user's request. Let s be a proposition reflecting the meaning of some given sentence. Cooper says that s is relevant to q if s is a necessary assumption one needs to make in order to prove q. Formally, it means that s belongs to a minimal set S of propositions that entail q:

$$rel(s,q) \iff \exists S : s \in S, S \models q, S - s \not\models q$$

A document D is considered relevant to the request if it contains at least one sentence s that is relevant to q. Cooper's definition is attractive in that it allows relevance to be defined on a sub-document level. This makes it reasonable to allow that documents may discuss different topics and that a single document may be relevant to very different requests. The definition also allows relevance to be extended into *novel* or *marginal* relevance, which will be discussed later in this section.

There are two main criticisms of Cooper's definition of relevance. One is operational, and has to do with the inherent difficulty of transforming natural language into logical propositions. The other is conceptual – Cooper's relevance is a binary relation between a query (expressed in logical form) and a document. The user is not in the picture, and neither is there any way to factor in the task that prompted the search in the first place. Consequently, with Cooper's definition a document ought to be judged relevant by every user that happens to generate q as their request, regardless of the task they are faced with or their personal preferences. An attempt to improve Cooper's definition was made by Wilson [150], who introduced the idea of *situational relevance*. Wilson's concept of relevance includes the situation in which the search is performed, user's goals, as well as the information already known to the user prior to examining a given document.

Cooper's definition of relevance experienced a strong revival when it was re-formulated in 1986 by Van Rijsbergen [138]. Van Rijsbergen's work was followed by a large number of publications describing relevance in terms of various formal logics [98, 99, 100, 101, 87, 122, 31, 32]. One important characteristic of most, if not all of these re-formulations is that they try to replace strict logical entailment ($d \models q$) with some weaker, but more tractable form. For example Van Rijsbergen [138, 139, 140] replaces the strict entailment with *logical implication* (denoted $d \rightarrow q$). Bruza [19, 20, 18] introduces a concept of *plausible entailment* and argues that a document d should be considered relevant as long as the query is at least plausibly entailed by some part of d. Lalmas and Van Rijsbergen [70, 71, 72] use modal logic and situation theory, treating a document as a *situation* and declaring it relevant if a flow of information from a document may lead to another situation, which in turn could strictly entail the query.

2.1.4 The binary nature of relevance

In most definitions relevance takes a form of a binary relation between a document and a query – a document is either relevant or it is not. A number of authors attempted to introduce graded notions of relevance, either in terms of discrete categories [38, 39, 65], or in terms of points or confidence intervals on the real line [42, 60]. While interesting in their own right, these definitions usually run into a number of practical difficulties. It turns out that graded relevance judgments are more costly to obtain, and the agreement between different judges is lower than for binary relevance. In addition, there is some evidence that human judges naturally prefer to use the end points of a given scale, hinting at the dichotomous nature of relevance [61]. Also, the methodology for evaluating retrieval effectiveness with graded relevance is not nearly as developed as it is for the case of binary relevance, where there is a small set of universally accepted performance measures.

Decision-theoretic view of relevance

When relevance is represented with real numbers, it is natural to start treating it as a measure of *utility* to a particular task, as was done by Cooper [28] and others. Zhai [157] has taken this view a step further, re-casting the entire retrieval process as that of minimizing the *risk* associated with missing usefull documents or presenting the user with garbage. In a noteworthy departure from related publications, Zhai's framework also incorporates a notion of a presentation strategy, where the same set of retrieved documents may result in different risk values, depending on how the documents are presented to the user.

2.1.5 Dependent and independent relevance

Most operational systems assume that relevance of a given document is independent of any other document already examined by the user, or of any other unexamined document in the collection. The assumption is motivated by a number of practical concerns. First, non-independent relevance judgments are considerably more expensive to obtain, since the judgment for a particular document will depend on the order in which all documents are presented to the judge. Second, retrieval algorithms themselves become computationally expensive when we have to search over the subsets of a large collection of documents. However, assuming independent relevance often takes us too far from the reality of an information seeking process, so a number of alternative definitions exist and will be outlined below.

Relevance and novelty

Redundancy of information is perhaps the most common reason for considering alternative, non-independent definitions of relevance. If a collection contains two near-duplicate documents the user is unlikely to be interested in reading both of them. They both may be topically relevant to the request, but once one of them is discovered, the second one may become entirely redundant, and irrelevant for information seeking purposes. To reflect the value of novelty, Carbonell and Goldstein [21] proposed the concept of *maximal marginal relevance* (MMR). Their idea was to provide a balance between the topical relevance of a document to the user's request, and redundancy of that document with respect to all documents already examined by the user. Allan and colleagues [6, 4] recognized the fact that novelty and redundancy must also be addressed on a sub-document level. For example, a document may be mostly redundant, but may contain a small amount of novel and very pertinent information. Something like this often happens in news reporting, where journalists tend to re-use a lot of previously reported information, interspersing it with important new developments. In Allan's formulation there are two separate definitions of relevance – topical relevance and novel relevance, and system performance is evaluated independently with respect to both of them. The importance of novelty in ranking has led to the establishment of the *Novelty Track* within the Text Retrieval Conference (TREC) [49], which has attracted a large number of publications in the last few years.

Relevance of a set of documents

If the information need of the user is sufficiently complex, it may be possible that no individual document completely satisfies that need by itself. However, information pooled from several documents may be sufficient. In this case the assumption of independent relevance clearly does not hold and we have to conceptualize relevance of a *set of documents* to the information need. For a single document, we may define a notion of *conditional* relevance, where conditioning is with respect to other documents included in the retrieved set. One of the first formal models of conditional relevance is due to Goffman [48], who defines the relevant set as a *communication chain* – a closed sequence of documents where relevance of a given document is conditioned on the previous document in the sequence (and must exceed a certain threshold for the document to be included). Goffman's model was evaluated by Croft and Van Rijsbergen [36] with results suggesting that the model, while considerably more expensive from a computational standpoint, was not superior to simpler forms of retrieval.

A major step away from independent document relevance was taken by Van Rijsbergen when he defined the now famous *cluster hypothesis*[137]:

> "Documents that cluster together tend to be relevant to the same requests."

The hypothesis was tested in a large number of studies, notably [62, 141, 144, 55, 82]. However, with few exceptions all these studies evaluate relevance at the level of individual documents in a cluster, rather than relevance of the cluster as a whole. Clustering was used primarily as a different way to organize retrieved documents.

Aspect relevance

In some cases, a complex information need can be broken down into smaller independent components. These components are often called *aspects*, and the goal of the retrieval system is to produce a set of documents that cover as many aspects of the overall need as possible. In this setting, it is common to introduce *aspect relevance*, which is topical relevance of a single document to a particular aspect of the overall need. Similarly, *aspect coverage* refers to the number of aspects for which relevant documents exist in the retrieved set. Aspect relevance and aspect coverage have been extensively studied in the Interactive Track of the TREC conference. A formal mechanism for modeling aspect utility was integrated by Zhai [157] into his risk-minimization framework.

2.2 Attempts to Construct a Unified Definition of Relevance

Faced with a growing number of definitions of relevance, several authors attempted to come up with a unified definition, which would classify and relate various notions of relevance. Some of the more prominent attempts were made by Saracevic [121], Mizarro [91, 92] and Ingwersen [30]. In this section we will briefly describe one particularly interesting proposal to formalize relevance by embedding it in a partially ordered four-dimensional space.

According to Mizarro [92], almost any reasonable definition of relevance can be represented as a vector consisting of four variables: *Information*, *Request*, *Time*, *Components*. The four dimensions of Mizarro's space have the following interpretations:

1. **Information type.** The first dimension of relevance is the kind of information resource for which we are defining relevance. In Mizarro's definition, this dimension can take one of three values: *document*, *surrogate* and *information.* Here *document* refers to the physical item a user will receive as the result of searching – the full text of a document, or, in the case of multi-media retrieval, a complete image, a full audio or video, file. *Surrogate* refers to a condensed representation of an information item, such as a list of keywords, an abstract, a title, or a caption. *Information* refers to changes in the user's state of knowledge as a result of reading or otherwise consuming the contents of a document. Note that information is a rather abstract concept, it depends on user's state of knowledge, his attentiveness, his capacity to comprehend the contents of the document and an array of other factors.
2. **Request type.** The second dimension of relevance specifies a level at which we are dealing with the user's problem. Mizarro defines four possible levels: *RIN*, *PIN*, *request* and *query.* The first (*RIN*) stands for Real Information Need and defines the information that will truly help the user solve the problem that prompted him to carry out a search in the first place. Needless to say, the user may not even be fully aware of what constitutes his real information need, instead he *perceives* it, and forms a mental image of it. That image is called *PIN*, or Perceived Information Need. Once the user knows (or rather thinks that he knows) what he is searching for, he formulates a *request.* A request is a natural language specification of what the user wants to find, something that might be given to a knowledgeable librarian or an expert in the field. A request is a way of communicating the *PIN* to another human being. Finally, this request has to be turned into a *query*, which is something that can be recognized by a search engine, perhaps a list of key words, a boolean expression or an SQL query. A query is a way of communicating the request to a machine.
3. **Time.** The third dimension of relevance reflects the fact that searching is not a one-shot process. The user may not see any relevant items in the initial retrieved set, and this may prompt him to re-formulate the *query*, perhaps changing the boolean structure, or adding some additional keywords. If the user does find relevant items, information in these items may prompt him to formulate different *requests*, or perhaps even force him to re-think what it is he wants to find, thus changing the perception of his information need (*PIN*). The real information need (*RIN*) stays constant and unchanging, since it is refers to what the user will ultimately be satisfied with. Mizarro endows the third dimension of relevance with a discrete progression of time points: $\{i_0, p_0, r_0, q_0, q_1, \ldots, r_1, q_{k+1}, \ldots, p_1, q_{m+1}, q_{n+1}, \ldots\}$. Here i_0 refers to the time the real information need (*RIN*) came to existence, p_0 is the time when the user perceived it, and decided what he wants to search for, r_0 is the time when he formulated a natural-language request, and q_0 is the time when that request turned into a query for a search engine. Proceeding further, q_1 is the time of the first re-formulation

of the request into a different query, r_1 is the first re-formulation of the natural-language request, and p_1 is the first time when the user changed the perception of what he wants to find. A request change r_i is always followed by one or more attempts to re-formulate the query q_j, and for every change in user's *PIN* there is at least one attempt to formulate a new request.

4. **Components.** The final dimension of relevance in Mizarro's framework specifies the nature of relationship between the first and the second dimension. It can be *topical* relevance, *task* relevance, *context* relevance, or any combination of the three. *Topical* relevance is concerned with semantic similarity in the content of the two items. *Task* relevance specifies that the item or information contained in it is useful for the task the user is performing. *Context* includes anything that is not covered by the topic and the task. Mizarro's *context* is a kind of "miscellaneous" category that subsumes the notions of novelty, comprehensibility, search cost, and everything else that does not seem to fit elsewhere in his formulation.

Mizarro [92] argues that his framework provides a useful tool for classifying and comparing various definitions of relevance. For example, the simple definition we provided in section 2.1.1 corresponds to *topical* relevance of a *surrogate* to the *query* at time q_0. To contrast that, Vickery's [142, 143] notion of "user relevance" relates the *information* in a document to the real information need (*RIN*) with respect to *topic*, *task* and *context* together.

At this point it is helpful to pause and admit that most practical retrieval models assume a fairly conservative definition of relevance restricted to a small portion of Mizarro's space. Most operational systems are concerned exclusively with *topical* relevance of full-text *documents* to natural-language *requests*. While there is active research on modeling *task* relevance, or taking *context* into account, the majority of experimental studies address only topicality. Any type of relevance with respect to perceived information need (*PIN*) is difficult to assess since there is only one user that knows what that *PIN* is. Furthermore, if we are dealing with *PIN*, we cannot cross-validate relevance judgments across different annotators. Modeling the real information need (*RIN*) is next to infeasible, since no one (not even the user) really knows what that is. Similarly, with very few exceptions (e.g. [81]) no one attempts to explicitly model the *information* contained in a given document – we simply do not have the frameworks that are simultaneously rich enough to represent arbitrary human discourse and robust enough to work on real-world data. On the other end of the spectrum, document *surrogates* and boolean *queries*, while quite popular in the past, are rarely used by the modern retrieval engines. Most current systems use full-text indexing, often with positional information and directly support natural-language requests.

2.2.1 Relevance in this book

For the scope of this book, we will concern ourselves with the popular view of relevance. In terms of Mizarro's classification, we will be constructing a formal model of relevance dealing with:

1. **Documents.** We will be operating on complete surface representations of information items (i.e. full text of documents, image bitmaps, segments of video, etc.)
2. **Requests.** We are concerned only with observable natural-language requests, for which we can obtain relevance judgments. However, our model will involve a notion similar to the real information need (**RIN**), which will play the role of a latent variable. We will use the words "query" and "request" interchangeably.
3. **Non-interactive.** We will not be modeling any evolution of user's information need. Our model explicitly accounts for the fact that a single information need can be expressed in multiple forms, but we do not view these in the context of an interactive search session.
4. **Topicality.** We will be interested exclusively in *topical* relevance, i.e. the semantic correspondence between a given request and a given document. We will not be addressing issues of presentation, novelty, or suitability to a particular task.

2.3 Existing Models of Relevance

This book is certainly not the first endeavor to treat relevance in probabilistic terms. Some of the more prominent examples are the 2-Poisson indexing model developed independently by Bookstein and Swanson [15, 16] and Harter [52], the probabilistic retrieval model of Robertson and Sparck Jones [117], the probabilistic flavors [123] of Van Rijsbergen's logical model [139], the inference-network model developed by Turtle and Croft [135, 134], the language modeling approach pioneered by Ponte and Croft [106, 105] and further developed by others [90, 56], and the recent risk minimization framework of Zhai and Lafferty [68, 157]. While we cannot provide a comprehensive review of all probabilistic models, we will devote the remainder of this chapter to a brief description of those models that had a particularly strong influence on the development of our generative model.

2.3.1 The Probability Ranking Principle

At the foundation of almost every probabilistic model of IR lies an intuitive principle that a good search system should present the documents in order of their probability of being relevant to the user's request. It appears that this idea was first formally stated by Cooper in a private communication to Robertson, who published it in the following form [114]:

> The Probability Ranking Principle (PRP): If a reference retrieval system's response to each request is a ranking of the documents in the collections in order of decreasing probability of usefulness to the user who submitted the request, where the probabilities are estimated as accurately as possible on the basis of whatever data has been made available to the system for this purpose, then the overall effectiveness of the system to its users will be the best that is obtainable on the basis of that data.

Symbolically, if D is used to denote every observable property of a given document in a search task, and if R is a binary variable indicating whether or not that document is relevant to the search request, then optimal performance would result from ranking documents in the decreasing order of $P(R=1|D)$. The words "optimal performance" refer both to informal user satisfaction (as long as redundancy is ignored) and to formal measures of evaluating IR performance. For example, it is not difficult to prove that using PRP to rank all documents in a given collection will lead to the greatest expected number of relevant hits in the top n ranks for every value of n (see [114]). Once that is established, it becomes obvious that the PRP maximizes *recall* and *precision* at rank n[1], as well as any measures that are derived from recall and precision, such as the F-measure [137], R-precision and mean average precision (MAP). Robertson [114] and Van Rijsbergen [137] also demonstrate that the Probability Ranking Principle leads to minimal decision-theoretic loss associated with retrieving a set of n top-ranked documents.

The probability ranking principle, as stated by Robertson is quite broad – it does not restrict us to any particular type of relevance, and the document representation D can be potentially very rich, covering not just the topical content of the document. In fact D could include features determining readability of the document, its relation to the user's preferences or suitability for a particular task. Similarly, R may well refer to "pertinence" or any other complex notion of relevance. The only restriction imposed by the PRP is that relevance of a particular document be scalar and independent of any other document in the collection. Consequently, PRP cannot handle issues of novelty and redundancy, or cases where two documents are relevant when put together, but irrelevant when viewed individually. Robertson [114] also cites a curious counter-example (due to Cooper) regarding the optimality of the principle. The counter-example considers the case when we are dealing with two classes of users who happen to issue the same request but consider different documents to be relevant to it. In that case PRP will only be optimal for the larger of the two user classes.

While PRP itself is quite general, in practice most probabilistic models take a somewhat more narrow view of it. In most cases the relevance R is

[1] *Recall* is defined as the number of relevant documents retrieved in the first n ranks, divided by the total number of relevant documents. *Precision* is the number of relevant documents in the first n ranks, divided by n.

restricted to mean *topical* relevance of a full-text document to a natural-language request. Relevance is also assumed to be fixed *a priori* in a form of relevance judgments for every document in the collection. These judgments are not directly observable by the system, but it is assumed that they exist and that they do not change in the course of a search session.

In the remainder of this section we will take a detailed look at several prominent probabilistic models of information retrieval. All of these models are either directly based on the PRP, or can be closely related to the principle. The major distinction between the models lies in how the authors choose to estimate the probability of relevance $P(R=1|D)$.

2.3.2 The Classical Probabilistic Model

We will first consider a probabilistic model of retrieval proposed by Robertson and Sparck Jones in [117]. The model was initially named the *Binary Independence Model*, reflecting the basic assumptions it made about occurrences of words in documents. However, since 1976 the model has been re-formulated a number of times to a degree where it can hardly be called "binary" and, as we shall argue later on, the term "independence" is also questionable. The model is also known as the *Okapi model*, the *City model*, or simply as *the probabilistic model*. A very detailed account of the recent state of the model is provided by the original authors in [131, 132]. What follows is our interpretation of the model. An attentive reader may find that our description is different in two ways from the way the model is presented by the authors [117, 131, 132]. First, we choose to describe the model in terms of probabilities, as opposed to log-likelihood weights. Second, we make explicit some of the assumptions that are never stated by the authors, particularly in section 2.3.2. Both steps are taken to make the description more compatible with subsequent material.

Robertson and Sparck Jones start the development of their model by transforming the probability ranking principle into the rank-equivalent likelihood ratio:

$$\begin{aligned} P(R=1|D) &\stackrel{\text{rank}}{=} \frac{P(R=1|D)}{P(R=0|D)} \\ &= \frac{P(D|R=1)P(R=1)}{P(D|R=0)P(R=0)} \\ &\stackrel{\text{rank}}{=} \frac{P(D|R=1)}{P(D|R=0)} \end{aligned} \tag{2.1}$$

Here R is a Bernoulli random variable indicating whether or not a given document is relevant and D is a random variable representing the contents of that document. We assume that D takes values in some finite space $\mathcal{D}$, and that P represents a joint probability measure over $\{0,1\} \times \mathcal{D}$. The first step of equation (2.1) comes from the fact that R is a binary variable and $\frac{P}{1-P}$ is a monotone (rank-preserving) transformation of P. The second step is a straightforward application of Bayes' rule. In the third step we observe that the factor $\frac{P(R=1)}{P(R=0)}$ is a constant independent of D, and thus does not affect the ranking of documents. In order to proceed further we need to specify the nature of the document space $\mathcal{D}$. The space has to be flexible enough to capture the semantic content of the documents, and yet simple enough to allow estimation from limited data. Robertson and Sparck Jones take $\mathcal{D}$ to be the space of all subsets of the vocabulary $\mathcal{V}$ of our collection. Equivalently, a document D is a binary occurrence vector, such that $D_v = 1$ if word number v is present in the document, otherwise $D_v = 0$. The document space $\mathcal{D}$ is then the space of all possible binary vectors over the vocabulary: $\mathcal{D} = \{0,1\}^{N_\mathcal{V}}$, and the entire probability space of the model in question is $\{0,1\} \times \{0,1\}^{N_\mathcal{V}}$ – the same finite space as the space of $(N_\mathcal{V}+1)$ coin tosses. As the next step in developing their model, Robertson and Sparck Jones assume that binary random variables D_i are mutually independent given the value of R, allowing them to factor the probabilities $P(D|R=1)$ and $P(D|R=0)$ as follows:

$$
\begin{aligned}
\frac{P(D{=}\mathbf{d}|R{=}1)}{P(D{=}\mathbf{d}|R{=}0)} &= \prod_{v\in\mathcal{V}} \frac{P(D_v{=}d_v|R{=}1)}{P(D_v{=}d_v|R{=}0)} \\
&= \left(\prod_{v\in D} \frac{p_v}{q_v}\right)\left(\prod_{v\notin D} \frac{1-p_v}{1-q_v}\right) \qquad (2.2)
\end{aligned}
$$

The first step in equation (2.2) comes from assuming independence between random variables D_v. We honor the common practice of using capital letters (e.g. D_v) to denote random variables and lowercase letters (e.g. d_v) to refer to their observed values. The second step comes from the fact that D_v can only take values of 0 and 1, and using a shorthand $p_v = P(D_v{=}1|R{=}1)$ and $q_v = P(D_v{=}1|R{=}0)$. Also note the slight abuse of notation in the product subscripts: expression $v{\in}D$ really means $\{v{\in}\mathcal{V} : d_v{=}1\}$. As the final step Robertson and Sparck Jones desire that an empty document ($\mathbf{0}$) correspond to a natural zero in the log-space of their ranking formula. They achieve this by dividing equation (2.2) by $\frac{P(\mathbf{0}|R=1)}{P(\mathbf{0}|R=0)}$. Since that quantity is independent of any document, dividing by it will not affect document ranking and will yield the following final ranking criterion:

$$
\begin{aligned}
P(R{=}1|D{=}\mathbf{d}) &\propto \frac{P(D{=}\mathbf{d}|R{=}1)}{P(D{=}\mathbf{d}|R{=}0)} \Big/ \frac{P(D{=}\mathbf{0}|R{=}1)}{P(D{=}\mathbf{0}|R{=}0)} \\
&= \prod_{v\in D} \frac{p_v(1-q_v)}{q_v(1-p_v)} \qquad (2.3)
\end{aligned}
$$

Parameter estimation with relevance information

Next comes the problem of estimation: in order to apply equation (2.3) to document retrieval we need to specify the quantities p_v and q_v, which reflect the probabilities of the word v occurring in a relevant and a non-relevant document respectively. Robertson and Sparck Jones start with the case where the relevance variable R is observable, that is for every document D=$\mathbf{d}$ in the collection they know whether it is relevant or not. If that is the case, a natural estimate of p_v is the proportion of relevant documents that contain word v, and similarly q_v is the proportion of non-relevant documents containing v. However, when R is fully observable, there is really no point in ranking: we could simply return the documents for which $R = 1$. A more realistic case is when R is partially observable, i.e. for some documents we know whether they are relevant or not, for others R is unknown. This is precisely the environment in Information Filtering, Topic Tracking or Relevance Feedback tasks. For that case Robertson and Sparck Jones adjust the relative frequencies of v in the relevant and non-relevant documents by adding a constant 0.5 to all counts:

$$p_v = \frac{N_{1,v} + 0.5}{N_1 + 0.5}$$
$$q_v = \frac{N_{0,v} + 0.5}{N_0 + 0.5} \quad (2.4)$$

Here N_1 is the number of known relevant documents, $N_{1,v}$ of them contain the word v. Similarly N_0 and $N_{0,v}$ reflect the total number of known non-relevant documents and how many of them contain v. The constant 0.5 in equation (2.4) serves two purposes: first, it ensures that we never get zero probabilities for any word v, and second, it serves as a crude form of smoothing (shrinkage), reducing the variance of estimates over possible sets of feedback documents.

Parameter estimation without relevance information

Until this point development of the Okapi model was quite straightforward. Unfortunately, it was based on the relevance variable R being at least partially observable, and that is simply not the case in a typical retrieval environment. In a typical scenario the only thing we have is the user's request Q, usually expressed as a short sentence or a small set of keywords. All our probability estimates have to be based on Q and on the collection as a whole, without knowing relevance and non-relevance of individual documents. To complicate matters further, Q is not even present in the original definition of the model (eqs. 2.1-2.3), it becomes necessary only when we have no way of observing the relevance variable R. Faced with these difficulties, Robertson and Sparck Jones make the following assumptions:

1. p_v=q_v if $v \notin Q$. When a word is not present in the query, it has an equal probability of occurring in the relevant and non-relevant documents. The effect of this assumption is that the product in equation (2.3) will only include words that occur both in the document and in the query, all other terms cancel out.

2. p_v=0.5 if $v \in Q$. If a word does occur in the query, it is equally likely to be present or absent in a relevant document. The assumption was originally proposed by Croft and Harper [34] and later re-formulated by Robertson and Walker [112]. The effect is that p_v and $(1 - p_v)$ cancel out in equation (2.3), leaving only $\frac{1-q_v}{q_v}$ under the product.
3. $q_v \propto N_v/N$. Probability of a word occurring in a non-relevant document can be approximated by its relative frequency in the entire collection. Here N is the total number of documents in the collection, N_v of them contain v. This approximation makes $\frac{1-q_v}{q_v}$ be proportional to the *inverse document frequency* (IDF) weight – a simple but devilishly effective heuristic introduced by Sparck Jones in [128].

Note that assumption (3) is quite reasonable: for a typical request only a small proportion of documents will be relevant, so collection-wide statistics are a good approximation to the non-relevant distribution. The same cannot be said for assumptions (1) and (2).

2-Poisson extension of the classical model

The original definition of the classical model deals exclusively with the binary representation of documents and queries, where a word is either present or not present in the document. However, empirical evidence suggests that the number of times a word is repeated within a document may be a strong indicator of relevance, and consequently the Okapi model was extended to include term frequency information. The first step in such extension is to expand the space $\mathcal{D}$ that is used to represent the documents. Previously, $\mathcal{D}$ was the set of all subsets of vocabulary $\{0,1\}^{N_\mathcal{V}}$. In order to handle frequencies, one can expand $\mathcal{D}$ to be $\{0,1,2\ldots\}^{N_\mathcal{V}}$. Now the ranking formula from equation (2.3) becomes:

$$P(R{=}1|D{=}\mathbf{d}) \propto \frac{P(D{=}\mathbf{d}|R{=}1)}{P(D{=}\mathbf{d}|R{=}0)} \Big/ \frac{P(D{=}\mathbf{0}|R{=}1)}{P(D{=}\mathbf{0}|R{=}0)}$$
$$= \prod_{v \in D} \frac{p_v(d_v) q_v(0)}{q_v(d_v) p_v(0)} \qquad (2.5)$$

Here d_v is the number of times word v was observed in the document. $p_v(d_v)$ is a shorthand for $P(D_v{=}d_v|R{=}1)$, the probability of seeing d_v occurrences of v in a relevant document, and $q_v(d_v)$ is the corresponding probability for the non-relevant document. Robertson and Sparck Jones [132] base their estimates of p_v and q_v on the 2-Poisson indexing model developed by Harter [52]. Harter's formalism revolves around a notion of *eliteness*, which was developed to model the behavior of a human indexer. Imagine a librarian who decides which keywords should be assigned to a given document. If he picks word v as a keyword for document d, then we say that d belongs to the *elite* class of v. Otherwise d belongs to the *non-elite* class. We would expect that documents in the elite class of v are likely to contain many repetitions of v, while in the non-elite class v would primarily occur by chance. Harter assumed that frequency of v in both classes follows a Poisson distribution, but that the mean is higher in the elite class. Under this assumption, the frequency of v in the collection as a whole would follow a mixture of two Poissons:

$$P(D_v{=}d_v) = P(E{=}1)\frac{e^{-\mu_{1,v}}\mu_{1,v}^{d_v}}{d_v!} + P(E{=}0)\frac{e^{-\mu_{0,v}}\mu_{0,v}^{d_v}}{d_v!} \tag{2.6}$$

Here E is a binary variable specifying whether D is in the elite set of v, $\mu_{1,v}$ is the mean frequency of v in the elite documents, and $\mu_{0,v}$ is the same for the non-elite set. Since we don't know which documents are elite for a given word, we need some way to estimate three parameters: $\mu_{1,v}$, $\mu_{0,v}$ and $P(E{=}1)$. Harter's solution was to fit equation (2.6) to the empirical distribution of v in the collection using the method of moments. But eliteness is not quite the same as thing as relevance, since eliteness is defined for single words and cannot be trivially generalized to multi-word requests. In order to fit Harter's model into the Okapi model the authors had to make some adjustments. Robertson and Walker [118] proposed to condition eliteness on R, and assumed that once we know eliteness, the frequency of v in a document is independent of relevance:

$$p_v(d_v) = P(E{=}1|R{=}1)\frac{e^{-\mu_{1,v}}\mu_{1,v}^{d_v}}{d_v!} + P(E{=}0|R{=}1)\frac{e^{-\mu_{0,v}}\mu_{0,v}^{d_v}}{d_v!} \tag{2.7}$$

$$q_v(d_v) = P(E{=}1|R{=}0)\frac{e^{-\mu_{1,v}}\mu_{1,v}^{d_v}}{d_v!} + P(E{=}0|R{=}0)\frac{e^{-\mu_{0,v}}\mu_{0,v}^{d_v}}{d_v!} \tag{2.8}$$

Substituting equations (2.7,2.8) back into the ranking formula (2.5), leads to a rather messy expression with a total of 4 parameters that need to be estimated for every word v: the mean frequencies in the elite and non-elite sets ($\mu_{1,v}$ and $\mu_{0,v}$), and the probability of eliteness given relevance or non-relevance ($P(E{=}1|R{=}1)$ and $P(E{=}1|R{=}0)$). This presents a rather daunting task in the absence of any relevance observations, leading the authors to abandon formal derivation and resort to a heuristic. They hypothesize that equations (2.7,2.8) might lead to the following term under the product in equation (2.5):

$$\frac{p_v(d_v)q_v(0)}{q_v(d_v)p_v(0)} \approx \exp\left\{\frac{d_v\cdot(1+k)}{d_v + k\cdot\left((1-b)+b\frac{n_d}{n_{avg}}\right)}\times\log\frac{N}{N_v}\right\} \tag{2.9}$$

The quantity under the exponent in equation 2.9 represents the well-known and highly successful *BM25* weighting formula. As before, d_v is the number of times v occurred in the document, n_d is the length of the document, n_{avg} is the average document length in the collection, N is the number of documents in the collection and N_v is the number of documents containing v. k and b represent constants that can be tuned to optimize performance of the model on the task at hand. We stress the fact that equation (2.9) is not a derived result and does not follow from any set of meaningful assumptions about the constituents of equations (2.7,2.8). *BM25* is a work of art, carefully engineered to combine the variables that were empirically found to influence retrieval performance: term frequency d_v, document length n_d and the inverse document frequency $\log\frac{N}{N_v}$. It is simple, flexible and very effective on a number of tasks (see [131, 132]). Unfortunately, it has no interpretation within the probabilistic model.

Modeling dependence in the classical model

The assumption of word independence in the classical model is a favorite target of linguistically sophisticated critics and aspiring graduate students. No other aspect of the formalism has drawn so much criticism and so many failed endeavors to improve the model[2]. Recall that the assumption states that individual words D_i in the document are mutually independent given the relevance variable R. The assumption is formalized in equation (2.2) for binary document representation and in equation (2.5) for the non-binary case. The assumption is intuitively wrong – we know that words in a language are not independent of each other: supposing that presence of the word "politics" tells us nothing about occurrence of "Washington" is clearly absurd. The popular perception is that the assumption of independence is a necessary evil, it is tolerated simply because without it we would have to estimate joint probabilities for vectors involving half a million random variables each (typical vocabulary size), and that is clearly intractable. Another popular perception is that there must be a way to partially model these dependencies, bringing the model closer to reality, and surely improving the retrieval performance.

One of the first attempts to relax the independence assumption is due to Van Rijsbergen [136]. The idea is to allow pairwise dependencies between words, such that for every word v there exists a *parent* word $\pi(v)$ which influences presence or absence of v. There is also a *root* word v_0 which has no parent. Dependencies form a spanning tree over the entire vocabulary, the structure of that tree can be induced automatically from a corpus by maximizing some objective function. Van Rijsbergen suggested using the aggregate mutual information over the branches ($\sum_v I(v,\pi(v))$) as the objective to maximize, other measures may work equally well. Once the structure $\pi(\cdot)$ is determined, we can replace the probabilities $P(D_v{=}d_v|R)$ in equations (2.2) and (2.5) with their conditional counterparts $P(D_v{=}d_v|D_{\pi(v)}{=}d_{\pi(v)},R)$. After re-arranging the indices v in the products to descend down the tree, we have a way to model relevance without assuming mutual independence.

[2] It is a personal observation that almost every mathematically inclined graduate student in Information Retrieval attempts to formulate some sort of a non-independent model of IR within the first two to three years of his or her studies. The vast majority of these attempts yield no improvements and remain unpublished.

Unfortunately, empirical evaluations [51, 50] of the new model suggest that by and large it performs no better than the original. When improvements were observed they were mostly attributed to *expanding* the query with additional words, rather than to a more accurate modeling of probabilities. Disappointing performance of complex models is often blamed on combinatorial explosion of the number of parameters. However, in Van Rijsergen's model the total number of parameters is only twice that of the original formulation: we replace p_v in equation (2.2) with $p_{v,0}$ and $p_{v,1}$, reflecting absence and presence of $\pi(v)$; the same is done for q_v. This suggests that number of parameters may not be the culprit behind the lack of improvement in retrieval accuracy. Neither can we blame performance on the particular choices made in [136] – during the two decades that passed, Van Rijsbergen's idea has been re-discovered and re-formulated by various researchers in wildly different ways [26, 27, 64, 80, 110, 137, 156]. In most cases the results are disappointing: consistent improvement is only reported for very selective heuristics (phrases, query expansion), which cannot be treated as formal models of word dependence. The pattern holds both when relevance is not observable (ad-hoc retrieval) and when there are a lot of relevant examples (text classification). In the latter case even phrases are of minimal value.

Why dependency models fail

It is natural to wonder why this is the case – the classical model contains an obviously incorrect assumption about the language, and yet most attempts to relax that assumption produce no consistent improvements whatsoever. In this section we will present a possible explanation. We are going to argue that the classical Binary Independence Model really *does not* assume word independence, and consequently that there is no benefit in trying to relax the non-existent assumption. Our explanation is an extension of a very important but almost universally ignored argument made by Cooper in [29]. Cooper argues that in order to arrive at equation (2.2), we only need to assume *linked dependence* between words D_i, and that assumption is substantially weaker than independence. Cooper's argument is as follows. Consider the case of a two-word vocabulary $\mathcal{V}=\{a,b\}$, and suppose we do not assume independence, so $P_1(D_a,D_b)$ is their joint distribution in the relevant class, $P_0(D_a,D_b)$ is the same for the non-relevant class. Now for a given document $D=\mathbf{d}$, consider the following quantities:

$$\begin{aligned} k_1 &= \frac{P_1(d_a,d_b)}{P_1(d_a)P_1(d_b)} \\ k_0 &= \frac{P_0(d_a,d_b)}{P_0(d_a)P_0(d_b)} \end{aligned} \tag{2.10}$$

By definition, k_1 is a measure of dependence between events $D_a{=}d_a$ and $D_b{=}d_b$ in the relevant class; it tells as how wrong we would be if we assumed

D_a to be independent of D_b. If $k_1{>}1$, the events d_a and d_b are positively correlated in the relevant class, $k_1{<}1$ means they are negatively correlated. k_0 plays the same role for the non-relevant class. Without assuming independence, the posterior odds of relevance (equation 2.1) takes the form:

$$\begin{aligned} P(R{=}1|D{=}\mathbf{d}) &\propto \frac{P(D{=}\mathbf{d}|R{=}1)}{P(D{=}\mathbf{d}|R{=}0)} \\ &= \frac{P_1(d_a,d_b)}{P_0(d_a,d_b)} \\ &= \frac{k_1 P_1(d_a)P_1(d_b)}{k_0 P_0(d_a)P_0(d_b)} \end{aligned} \tag{2.11}$$

When Robertson and Sparck Jones [117] assume that D_a and D_b are independent, they in effect assume that $k_1{=}1$ and $k_0{=}1$. But Cooper [29] correctly points out that to justify equation (2.2) we only need to assume $k_1{=}k_0$, which is much less restrictive: k_1 and k_2 can equal any number, not just 1. This is Cooper's *linked dependence* assumption, it demonstrates that the classical model actually allows for any degree of dependence between words a an b, as long as that dependence is exactly the same in the relevant and non-relevant classes.

Cooper's assumption is certainly more reasonable than mutual independence, but it has its limitations. For example, if the user's request happens to deal with compound concepts, such as "machine translation", it would be disastrous to assume the same degree of dependence for these two words in the relevant and non-relevant documents. Additionally, linked dependence presented by Cooper, becomes more and more restrictive as we consider larger vocabularies and deal with factors k_1 and k_0 of the form $\frac{P(d_1 \ldots d_n)}{P(d_1)\ldots P(d_n)}$. However, we would like to argue that Cooper's argument can be taken one step further, yielding an even weaker assumption that can withstand the counterexample given above. We will refer to this as the assumption of **proportional interdependence.**Let $\mathcal{V}$ be a general vocabulary. As a first step, we will restrict our discussion to the simple case where only first-order dependencies exist between the words: a word v may only depend on one other word, as in Van Rijsbergen's model [136]. We will go a step further and allow each word v to have potentially different parents $\pi_1(v)$ and $\pi_0(v)$ in the relevant and non-relevant dependence trees. We know that under a first-order model, the joint distribution $P(D{=}\mathbf{d}|R{=}r)$ decomposes into a product of conditional probabilities $P(D_v{=}d_v|D_{\pi_r(v)}{=}d_{\pi_r(v)}, R{=}r)$, one for each word v in the vocabulary. Inspired by Cooper, we define the factor $k_{v,r}$ to be the ratio of the conditional probability to the unconditional one:

$$k_{v,r} = \frac{P(D_v{=}d_v|D_{\pi_r(v)}{=}d_{\pi_r(v)}, R{=}r)}{P(D_v{=}d_v|R{=}r)} \tag{2.12}$$

Now the version of equation (2.1) appropriate for a first-order dependence model will take the following form:

$$P(R{=}1|D{=}\mathbf{d}) \propto \prod_{v \in \mathcal{V}} \frac{P(D_v{=}d_v|D_{\pi_1(v)}{=}d_{\pi_1(v)}, R{=}1)}{P(D_v{=}d_v|D_{\pi_0(v)}{=}d_{\pi_0(v)}, R{=}0)}$$
$$= \prod_{v \in \mathcal{V}} \frac{P(D_v{=}d_v|R{=}1)}{P(D_v{=}d_v|R{=}0)} \cdot \frac{k_{v,1}}{k_{v,0}} \tag{2.13}$$

Equation (2.13) makes it very clear that the first-order model is rank-equivalent to the independence model if and only if $\prod_v \frac{k_{v,1}}{k_{v,0}}$ is a constant independent of $\mathbf{d}$. An equivalent statement is that cumulative pairwise mutual information between the presence of a randomly picked word and the presence of its parent differs by a constant k (independent of $\mathbf{d}$) in the relevant and non-relevant classes:

$$\sum_{v \in \mathcal{V}} \log \frac{P_1(d_v, d_{\pi_1(v)})}{P_1(d_v)P_1(d_{\pi_1(v)})} = k \sum_{v \in \mathcal{V}} \log \frac{P_0(d_v, d_{\pi_0(v)})}{P_0(d_v)P_0(d_{\pi_0(v)})} \tag{2.14}$$

Informally, equation (2.14) means that *on average*, all the words in a given document have about as much interdependence under the relevant class (P_1) as they do under the non-relevant class (P_0). The key phrase here is "on average": equation (2.14) does not require that any two words be equally dependent under P_1 and P_1 – that is precisely Cooper's linked dependence. Instead, equation (2.14) allows some words to be strongly dependent only in the relevant class (e.g., "machine" and "translation"), as long as on average they are balanced out by some dependencies that are stronger in the non-relevant class. They don't even have to balance out exactly ($k{=}0$), the only requirement is that whatever disbalance exists be constant across all documents.

We view the above as a strong result: the independence model is equivalent to any first-order dependence model under a very weak **proportional interdependence** assumption, that we personally believe holds in most situations.[3] Indeed, we see no reason to believe that an arbitrary set of documents in the collection (the relevant set for some request) will exhibit a stronger cumulative dependence over all words than will the complement of that set. The meaning of this result is that any attempt to replace independence with first-order dependence is very likely to produce no improvements, other than by

[3] If desired, one could certainly test the empirical validity of the proportional interedependence assumption for a given collection. The test would proceed as follows. **(1)** partition the collection into relevant and non-relevant sets using complete relevance judgments. **(2)** estimate the dependency structures $\pi_1()$ and $\pi_0()$ for the relevant and non-relevant classes (e.g., using Van Rijsbergen's method). **(3)** construct maximum-likelihood estimates for conditional distributions $P_r(v|\pi_r(v)) : r{\in}\{0,1\}$. **(4)** compute the value k in equation (2.14) for each document d in the collection. **(5)** perform a statistical goodness-of-fit test, comparing the set of values k observed for the relevant documents against the values observed for non-relevant documents. If the null hypothesis (that the populations are identical) cannot be rejected, then the proportional interdependence assumption holds for this collection.

accident. We also point out that this result may not be limited to first-order dependencies. One could define factors $k_{v,0}$ and $k_{v,1}$ for higher-order models where word probabilities are conditioned on *neighborhoods* $\eta(v)$ instead of of *parents* $\pi(v)$. Admittedly, the argument becomes somewhat more elaborate; we have not worked out the details. As a conclusion to this section, we would like to stress the following:

> Contrary to popular belief, word independence is not a necessary assumption in the classical probabilistic model of IR. A necessary and sufficient condition is proportional interdependence, which we believe holds in most retrieval settings. If there is anything wrong with the classical model, it is not independence but the assumptions made in the estimation process (see sections 2.3.2 and 2.3.2).

2.3.3 The Language Modeling Framework

We will now turn our attention to a very different approach to relevance – one based on statistical models of natural language. Statistical language modeling is a mature field with a wide range of successful applications, such as discourse generation, automatic speech recognition and statistical machine translation. However, using language modeling in the field of Information Retrieval is a relatively novel development. The approach was proposed by Ponte and Croft [106, 105] in 1998, and in the short time since then it has attracted a tremendous level of interest and a growing number of publications each year. In this section we will outline the original language modeling approach to IR [106] and briefly mention some of the more prominent extensions.

One of the main motivations Ponte and Croft had for developing the language modeling approach was to get away from the heuristics that came to dominate the probabilistic model of Robertson and Sparck Jones [117]. Recall that heuristics in the classical model arise when we are given no examples to estimate the probabilities p_v associated with relevant documents. Ponte and Croft's solution to this problem was quite radical – it was to remove the explicit relevance variable R, and construct a probabilistic model around the document and the user's query. The authors hypothesized that for every document $D{=}\mathbf{d}$, there exists an underlying language model M_d. Now, if the query $Q{=}\mathbf{q}$ looks like it might be a random sample from M_d, we have a reason to believe that $\mathbf{d}$ is relevant to $\mathbf{q}$. Informally, we can think of M_d as a crude model reflecting the state of mind of the author who created document $\mathbf{d}$. If the same state of mind is likely to produce the query $\mathbf{q}$, then it is likely that $\mathbf{q}$ is topically related to $\mathbf{d}$, hence $\mathbf{d}$ would be topically relevant.

The effect of Ponte and Croft's argument is that they could replace the probability of relevance $P(R{=}1|D{=}\mathbf{d})$ with the probability of observing the query from the language model of the document $P(Q{=}\mathbf{q}|M_d)$. This is a crucial step: it allows the authors to avoid the uncertainty associated with the unobserved relevance variable R. Indeed, Q is observable, and there exists a substantial body of statistical literature to help us in estimating M_d from the observed document **d**. Retrieval in Ponte and Croft's model can be decomposed into two steps. First, we have to use the observation **d** to construct our estimate of the underlying document language model M_d. Second, we can compute the probability $P(Q{=}\mathbf{q}|M_d)$ of observing **q** as a random sample from M_d, and rank all documents in the decreasing order of that probability.

Multiple-Bernoulli language models

Ponte and Croft represent queries in the same space that was used by Robertson and Sparck Jones in the Binary Independence Model. If $\mathcal{V}$ is a vocabulary of $N_\mathcal{V}$ words, the query space $\mathcal{Q}$ is the set of all subsets of vocabulary ($\{0,1\}^{N_\mathcal{V}}$). The query Q is a vector of $N_\mathcal{V}$ binary variables Q_v, one for each word v in the vocabulary. The components Q_i are assumed to be mutually independent conditioned on the language model M_d. The language model itself is a vector of $N_\mathcal{V}$ probabilities $p_{d,v}$, one for each word v. The probability of observing a query $Q{=}\mathbf{q}$ from a given model $M_d{=}\mathbf{p}_d$ is:

$$
\begin{aligned}
P(Q{=}\mathbf{q}|M_d{=}\mathbf{p}_d) &= \prod_{v\in\mathcal{V}} P(Q_v{=}q_v|M_d{=}\mathbf{p}_d) \\
&= \prod_{v\in Q} p_{d,v} \times \prod_{v\notin Q} (1-p_{d,v})
\end{aligned}
\tag{2.15}
$$

Here again, $v{\in}Q$ is a shorthand for $\{v{\in}\mathcal{V} : q_v{=}1\}$, and likewise for the complement set. Ponte and Croft propose the following way to compute $\mathbf{p}_d$ from the document **d**:

$$
p_{v,d} = \begin{cases} \left(\frac{d_v}{|\mathbf{d}|}\right)^{(1-r)} \left(\frac{1}{N_v}\sum_{\mathbf{d}'}\frac{d'_v}{|\mathbf{d}'|}\right)^{r} & \text{if } d_v > 0 \\ \left(\sum_{\mathbf{d}'} d'_v\right) / \left(\sum_{\mathbf{d}'} |\mathbf{d}'|\right) & \text{otherwise} \end{cases}
\tag{2.16}
$$

Here d_v is the number of times word v occurs in document **d**, $|\mathbf{d}|$ denotes the length of document **d**, N_v is the number of documents containing v and the summations go over every document $\mathbf{d}'$ in the collection. If a word v does not occur in the document, Ponte and Croft use its relative frequency in the entire collection. If a word does occur, the estimate is a weighted geometric average between the relative frequency in the document and the average relative frequency over all documents containing v. The weight is given by the parameter r, which according to the authors plays the role of *Bayesian risk*.

A probabilist will undoubtedly notice an inconsistency between equations (2.15) and (2.16). The former represents a *multiple-Bernoulli* distribution over the binary event space, but the probability estimates in equation (2.16) are based on non-binary frequencies d_v and would naturally arise if we assumed that **d** was a random sample from a *multinomial* distribution. Ponte and Croft never address the issue, but elsewhere [88] we show that the model can be made consistent by assuming that each document **d** is represented not by a single set of words, but by a set of $|\mathbf{d}|$ singleton sets, each assummed to be independently drawn from M_d.

Multinomial language models

As we mentioned above, term frequencies are somewhat unnatural in Ponte and Croft's model. The model is explicitly geared to capture the presence or absence of words, and does not recognize the fact that words can be repeated in the query. This is perfectly reasonable for short 2-3 word queries that are typical of web searches, but it is not a good assumption for the general retrieval setting. In order to take account of frequencies researchers have had to assume a different event space. Virtually every publication concerning language modeling in IR [126, 127, 90, 11, 152, 58, 56, 158, 159] presumes the following representation, though it is rarely stated in formal terms. Assume $\mathcal{V}$ is a vocabulary of $N_{\mathcal{V}}$ distinct words. Both documents and queries are viewed as strings (sequences) over $\mathcal{V}$. A document D of length m is a sequence of m random variables D_i, each taking values in the vocabulary $\mathcal{V}$. The query Q has the same representation: $Q{=}n, Q_1 \ldots Q_n$, such that $Q_i{\in}\mathcal{V}$ for each $i = 1 \ldots n$. The probability space for both documents and queries is the space of all possible sequences of words: $\mathcal{Q} = \mathcal{D} = I\!N \times \cup_{n=1}^{\infty} \mathcal{V}^n$. Note: most authors omit sequence length N from the representation. We make it explicit to define a single probability measure for strings of any length. Individual words Q_i in the sequence are assumed to be independent of N, independent of each other and identically distributed according to the language model M_d. M_d now plays the role of a discrete distribution over the vocabulary; its values are vectors of $N_{\mathcal{V}}$ probabilities, one for each word v: $\mathbf{p}_d{\in}[0,1]^{N_{\mathcal{V}}}$ such that $1{=}\sum_v \mathbf{p}_{d,v}$. The probability mass assigned by a language model M_d to some string Q is:

$$
\begin{aligned}
P(Q{=}\mathbf{q}|M_d{=}\mathbf{p}_d) &= P(N{=}n)\prod_{i=1}^{n} P(Q_i{=}q_i|M_d{=}\mathbf{p}_d) \\
&= \pi_n \prod_{i=1}^{n} p_{d,q_i}
\end{aligned}
\tag{2.17}
$$

Here $P(N{=}n) = \pi_n$ is some discrete distribution over string lengths; it is independent of everything else and is usually assumed to be uniform until some maximum length M and zero beyond that. A common way to estimate the language model is to assume that the document $\mathbf{d}$ itself represents a random sample drawn from M_d, and use relative frequencies of words in $\mathbf{d}$ as a maximum likelihood estimate $\mathbf{p}_d$. However, maximum likelihood estimation will naturally lead to zeros in the estimate, so some form of smoothing is required. From the large pool of available smoothing techniques [23, 158], most authors pick some form of linear interpolation between the maximum likelihood estimate and the "background" frequency of a word computed over the whole collection:

$$
p_{d,v} = \lambda_d \frac{n_{d,v}}{n_d} + (1-\lambda_d)\frac{n_{c,v}}{n_c} \tag{2.18}
$$

Here $n_{d,v}$ refers to the number of times the word v occurs in document $\mathbf{d}$, n_d is the length of $\mathbf{d}$, $n_{c,v}$ is the frequency of v in the entire collection and

n_c is the total number of words in the collection. Parameter λ_d is used to control the degree of variance in the estimator. Lafferty and Zhai [68] show that equation (2.18) is a natural Bayesian estimate that follows from assuming a Dirichlet prior with parameters proportional to $\frac{n_{c,v}}{n_c}$ over the simplex of all possible language models.

Multinomial and multiple-Bernoulli event spaces

As we already mentioned, the original language modeling framework proposed by Ponte and Croft [106] is defined over a binary event space, and does not allow for word repetitions. The multinomial approach described in section 2.3.3 does allow word frequencies, but that is not the only difference between the two frameworks. A much more important, and commonly overlooked difference is that the two approaches have very different and incompatible event spaces. The random variable Q means two completely different things in equations (2.15) and (2.17), and the meanings cannot be used interchangeably, or mixed together as was done in several publications. In Ponte and Croft's model, Q is a *vector* of binary variables Q_v. Each Q_v represents a *word* in the vocabulary, possible values of Q_v are 0 and 1 (absent and present). In the multinomial framework, Q is a sequence of variables Q_i, each Q_i represents a *position* in the query, and the values of Q_i are words. Note that the latter representation is absolutely not a requirement if we just want to model counts. We could have simply extended the range of Q_v in the Ponte and Croft model to include counts, as was done in the 2-Poisson extension of the Binary Independence Model. Doing this would give us a vector representation that is very different from the sequence representation described in section 2.3.3. To be specific, in the vector representation we have half a million random variables, each with two possible values: absent or present. In the sequence we effectively have only one variable (since Q_i are i.i.d.), but that variable can take half a million possible values.

Which of these representations is more suitable for Information Retrieval is a very interesting open question. Our feeling is that vector representation might be more natural. It allows us to estimate a separate distribution for every vocabulary word, makes no a-priori assumptions about word counts and allows us to explicitly model dependencies between different vocabulary words (though in light of section 2.3.2 we might wonder if dependencies are of any use at all). On the other hand, the sequence representation makes it more natural to model word proximity, phrases and other surface characteristics of text. In practice, the question of representation is all but settled – nearly every publication assumes sequences rather than vectors. This choice is largely a matter of consistency – in the fields adjacent to IR, such as Speech Recognition (ASR), Machine Translation (MT) and Natural Language Processing (NLP), language modeling is always concerned with sequences. In addition, a large body of language modeling publications in these fields serves as a gold-mine of estimation techniques that can be applied in IR – anything from n-gram and cache models in ASR, to translation models in MT, to grammars in NLP. For the remainder of this book, when we speak of language models we will refer to sequence models, as defined in section 2.3.3.

Independence in the language modeling framework

Just like the classical probabilistic model, the language modeling approach relies on making a very strong assumption of independence between individual words. However, the meaning of this assumption is quite different in the two frameworks. In the classical model, independence means that presence of "politics" does not affect presence of "Washington". In the language modeling framework, the independence assumption means that the identity of the n'th word in the sequence does not depend on any preceding or following words. The second assumption implies the first: "Washington" in position n is still not affected by "politics" in position m. However the converse is not true: in addition to the above, the language modeling framework assumes that "politics" in position m does not affect "politics" in position n, so in a sense a word is independent of itself. This is certainly not the case in the classical model, although assuming a Poisson distribution in section 2.3.2 is essentially equivalent, since Poisson sampling is a memory-less process.

Given that the assumption of independence is even stronger in LM than in the classical model, it should come as no surprise that several researchers attempted to relax the assumption. One of the first attempts is due to Song and Croft [126, 127]. In that work each query word Q_i was conditioned on the immediately preceding Q_{i-1}, forming a first-order Markov Chain. The same assumption was made about the documents, and consequently the language model M_d takes the form of a bigram model. The parameters were estimated using bigram frequencies in the document with back-off to the unigram and to collection-wide counts. Note that in this case one does experience a combinatorial explosion: the number of parameters is squared, not just doubled as in Van Rijsbergen's dependence model [136]. As might be expected, the new model did not yield consistent improvements over the original formulation. A similar model was proposed by Miller and colleagues [89, 90], but bigram performance was never evaluated. A very different formalism was recently attempted by Nallapati [94, 95], who tried to combine Van Rijsbergen's dependence model [136] with the multinomial model in section (2.3.3). The results were inconsistent. We are also aware of several unpublished studies, where different attempts to introduce dependencies between random variables Q_i or D_i did not lead to any improvements over the unigram model.

Faced with the poor performance of dependency models, can we repeat the argument made in section 2.3.2 and show that document ranking is not affected by the assumption of independence? Our intuition is that we cannot: the classical probabilistic model involved only two probability distributions (relevant and non-relevant), while in the language-modeling framework we are faced with a distinct probability distribution M_d for each document **d** in the collection. Furthermore, interdependencies accumulate only over the query words, not over the entire vocabulary as was the case with the classical model. We see no reason to believe that aggregate dependence among the query words will not heavily depend on M_d. However, we will provide an informal argument for why modeling dependence does not seem to help in IR, whereas it is absolutely essential in other fields that use language models (ASR, MT, NLP). The primary use of language models in fields other than IR is to ensure surface consistency, or well-formedness of strings of words. As an example, consider a speech recognition system, which typically consists of an acoustic model and a language model. To an acoustic model the utterance "I see defeat" may appear no different from a nonsensical "icy the feet". But any decent bigram model would favor the first string as more consistent. Similarly, in NLP a discourse generation system may use a grammar of English to translate a template-based representation of some action into a well-formed sentence. In these fields it is absolutely necessary to worry about the surface form of strings because the goal is to generate *new* strings of text in response to some input. If the system generates gibberish, it is useless. Information retrieval is different in the sense that it deals with *existing* strings of text, which are already well-formed. When we directly adopt an n-gram model from speech recognition or a maximum-entropy model from MT, we are in effect adopting a proven solution for a problem that we do not face. At the same time, the added complexity of the new model will likely translate to less reliable parameter estimates than the ones we had with the simpler model.

The above argument should be taken with a grain of salt. We are not suggesting that it is impossible to improve the dependency structure of the language modeling approach. We only claim that no improvement should be expected from *direct* models of dependence – models where random variables Q_i are *directly* conditioned on some $Q_{j \neq i}$. That, however, does not mean that no improvement will result from models that capture dependencies *indirectly*, perhaps via a hidden variable. For example, Berger and Lafferty [11] proposed to model information retrieval as statistical translation of a document into the query. In the simplest instantiation of their approach (model 1), latent words T_i are randomly sampled from the document model M_d, and then probabilistically *translated* into query words Q_i according to some distribution $P(Q_i|T_i)$. Under the translation model, the document ranking criterion (equation 2.17) takes the following form:

$$P(Q{=}\mathbf{q}|M_d{=}\mathbf{p}_d) \propto \prod_{i=1}^{n} \sum_{v \in \mathcal{V}} P(Q_i{=}q_i|T_i{=}v)P(T_i{=}v|M_d{=}\mathbf{p}_d)$$
$$= \prod_{i=1}^{n} \sum_{v \in \mathcal{V}} t_{v,q_i} p_{d,v} \qquad (2.19)$$

In the translation model there is no direct dependency between the query words Q_i or the document words D_i. Instead, the translation matrix $t_{v,q}$ provides a useful way to handle dependencies between *identities* of the words in the document and in the query. For instance, the translation model would correctly model dependency between the word "cat" in a document and the word "feline" in the query, if the translation matrix $t_{v,q}$ was estimated to reflect synonymy. The translation model was originally proposed as a general-purpose model of IR, but it found its greatest application in the field of cross-language retrieval, where documents in one language are queried using some other language. Two other prominent examples of the indirect approach to word dependencies are the latent aspect model of Hoffman [58], and the Markov-chain model proposed by Lafferty and Zhai [68].

2.3.4 Contrasting the Classical Model and Language Models

In the previous section we described two probabilistic frameworks for modeling relevance in information retrieval. One grew out of the Binary Independence Model, proposed by Robertson and Sparck Jones [117]; the other represents various developments of the language-modeling approach pioneered by Ponte and Croft [106]. This section will be devoted to looking at the benefits and drawbacks of the two frameworks and will serve as a bridge leading into the development of our own generative model of relevance.

Despite their fundamental differences, there are a number of similarities between the classical model and language models. Both approaches started with a focus on binary presence or absence of words, and, curiously, used exactly the same event space. Both rapidly evolved from binary occurrence to modeling word frequencies – one explicitly, via a Poisson distribution over the counts, the other, implicitly, by adopting a multinomial distribution over the vocabulary. Both frameworks appear to make a very strong assumption of independence, though we have argued that the meaning of *independence* is quite different in the two models. In the former, independence concerns word *identities* (presence of "politics" unaffected by "Washington"); in the latter word *positions* are assumed independent (first query/document word does not affect second word). Since both assumptions appear obviously incorrect, a lot of effort went into improving performance by modeling dependencies. Unfortunately, explicit models of dependence did not lead to consistent performance improvements in either framework. In an attempt to understand this curious effect, we provided two different arguments for why dependency models do not help in the classical framework and the language modeling framework. For the classical case we extended an argument initiated by Cooper [29] and showed that the model really does not assume independence, it is based on a much weaker assumption of proportional interdependence (see section 2.3.2 for details). For the language modeling framework we informally argued that explicit models of dependence will capture nothing but the surface form (well-formedness) of text, which has little to do with the topical content.

Relevance in the two frameworks

The point where language models become very different from the classical models is on the issue of relevance. The classical probabilistic model is centered around relevance: Robertson and Sparck Jones [117] start the development of their model directly from the probability ranking principle and proceed formally as far as they can (as long as relevance is observable). To contrast that, in the language modeling approach there is no explicit concept of relevance. Ponte and Croft [106] replace it with a simple generative formalism: probability of relevance $P(R|D)$ is assumed to be proportional to the probability of randomly drawing the query Q from the document model M_d. There is a clear benefit to this assumption: since both the query and the document are always fully observable, the model does not have to deal with the ambiguous concept of relevance. Ponte and Croft effectively turned a retrieval problem into an estimation problem. Instead of trying to model relevance we look for the best way to estimate the language model M_d for a given document d. This estimation step can be carried out in a systematic fashion, without resorting to heuristics that become necessary in the classical model.

However, absence of relevance raises the question of what to do in the rare cases when we do have relevant examples. To elaborate, suppose we have a small set of documents that are known to be relevant to a given query. How could we make use of this information to improve ranking of subsequent documents? The process is very straightforward in the classical probabilistic model – we simply use relative frequencies of words to get better probability estimates; the exact formulae are given in section 2.3.2. This can be done with both positive and negative (non-relevant) examples. With the language modeling framework incorporating relevance judgments is not nearly as clear. We cannot update probability estimates because there is no distribution associated with the relevant class. Updating the document model M_d makes no sense, since examples are relevant to the query, not to the document. Ponte [105] suggests that the only thing we can do is re-formulate the query, expanding it with words selected from relevant examples according to a heuristic weighting formula. Interestingly, language models allow for a very different kind of feedback that cannot be handled within the classical model. If for a given document **d** we have examples of relevant queries (queries for which **d** was judged relevant), we can certainly make use of those queries in adjusting the language model M_d. This form of feedback has recently been studied in [96].

Formal absence of relevance from the language modeling approach has also led to continued criticism of the framework [129, 111, 130]. Quoting from [130]: *"a retrieval model that does not mention relevance appears paradoxical"*. Responding to this criticism, Lafferty and Zhai [69] claim that the language modeling approach can be re-formulated to include a concept of relevance, albeit implicitly. To support their claim, Lafferty and Zhai argue that from a high-level viewpoint both frameworks operate with three random variables:

the query Q, the document D and the relevance variable R. Both frameworks attempt to approximate the posterior distribution over R, but the factorization of dependencies is done in different ways. In the classical probabilistic model, D is factored out first and conditioned on R and Q, leading to the familiar development:

$$\frac{P(R{=}1|D,Q)}{P(R{=}0|D,Q)} = \frac{P(D|R{=}1,Q)}{P(D|R{=}0,Q)} \cdot \frac{P(R{=}1,Q)}{P(R{=}0,Q)} \propto \frac{P(D|R{=}1,Q)}{P(D|R{=}0,Q)} \tag{2.20}$$

The leftmost portion of Figure 2.1 shows a graphical diagram of the dependencies implied by equation (2.20). Following convention, we use shaded circles to represent observed variables D and Q. A corresponding dependency diagram for the language modeling approach is shown in the middle of Figure 2.1. We use dashed lines to indicate that R is introduced somewhat artificially. The diagram results from factoring Q conditioned on R and D as follows:

$$\frac{P(R{=}1|D,Q)}{P(R{=}0|D,Q)} = \frac{P(Q|R{=}1,D)}{P(Q|R{=}0,D)} \cdot \frac{P(R{=}1,D)}{P(R{=}0,D)} \propto P(Q|R{=}1,D) \cdot \frac{P(R{=}1,D)}{P(R{=}0,D)} \tag{2.21}$$

In order to justify the second step above, Lafferty and Zhai have to assume that Q is independent of D if $R{=}0$, which means that the denominator of the first ratio does not affect the ranking and can be omitted. However, equation (2.21) still includes the ratio of joint probabilities over R and D, which is not present in the language modeling approach. To get rid of it, Lafferty and Zhai proposed to make R independent of D. This would leave $P(Q|R{=}1,D)$ as the only term that affects the ranking of documents, thus explaining the language modeling framework.

As a final note, we would like to suggest that there is a third way of factoring the joint distribution over R, D and Q. We could assume that the query Q and the document D are conditioned on the relevance variable R, and that Q and D are independent given R. This factorization is shown in the rightmost diagram in Figure 2.1, it forms the foundation for the model proposed in this book and will be discussed in great detail in the following chapter.

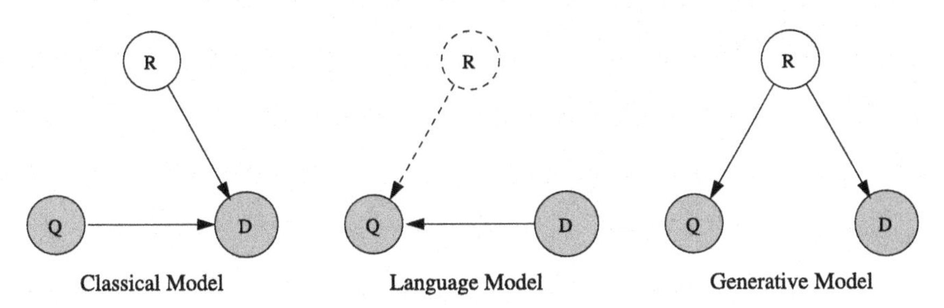

Fig. 2.1. Graphical diagrams showing dependencies between the query Q, the document D and relevance R variables in different probabilistic models of IR. Left: classical probabilistic model [117]. Middle: language modeling framework [106] according to [69]. Right: the generative model proposed in this book. Shaded circles represent observable variables.

3

A Generative View of Relevance

In this chapter we will introduce the central idea of our work – the idea that relevance can be viewed as a stochastic process underlying both information items and user requests. We will start our discussion at a relatively high level and gradually add specifics, as we develop the model in a top-down fashion.

3.1 An Informal Introduction to the Model

Suppose we find ourselves in the following environment. We have a user, or a group of users, performing a particular task. In order to perform that task effectively, the user needs to have access to some specific information. We also have a collection of information items, perhaps in textual form, perhaps as audio files, images, videos, or as a combination of these forms. We will refer to these items as *documents*. We assume that a subset of documents in this collection are *topically relevant* to the user's information need, which means that if the user examined these documents, he would judge them to discuss the subject matter that is directly related to his need. These documents form the *relevant* subset of the collection. It is also natural to allow the user to express his perception of the information need. We will interchangeably use the words *request* and *query* to refer to these expressions. We recognize the fact that the user may generate multiple requests for the same information need. We will place no restriction on the number or form of these requests, other than that they be consistent. By consistency we mean that the user looking at a given request should be able to determine whether he could have generated that request for his current information need. Assuming consistency allows us to talk about the *relevant query set*, i.e. the set that contains all requests that could have been produced by our user in response to his information need. With the above definitions, we are prepared to state the major assumption that will serve as the foundation for most of the work presented in this book. We choose to formulate this assumption in the form of a hypothesis:

The Generative Relevance Hypothesis (GRH): for a given information need, queries expressing that need and documents relevant to that need can be viewed as random samples from the same underlying generative model.

In effect what we are stating is that documents in the relevant set are all alike, in the sense that they can be described by some unknown generative process associated with the user's information need. Furthermore, any request the user might formulate to express his information need is expected to follow the same generative process. We will use the term **relevance model** to refer to a particular restriction of this generative process. The primary focus of this book is on methods for estimating relevance models and for using them to find the items relevant to the user's information need.

At this point we expect to encounter a lot of skepticism from a savvy IR practitioner. The generative hypothesis seems to imply that documents and queries are identically distributed, so a document relevant to a given query might look a lot like that query. That, of course, is an unreasonable assumption. For instance, if we are dealing with a web-searching scenario we expect a typical query to be a set of two, at most three keywords carrying no syntactic structure whatsoever. On the other hand, a relevant document is expected to be a complete webpage, hopefully containing more than the three words that the user typed into the search box. We expect the text in the webpage to be comprehensible, with well-formed grammar and a lot of "glue" words that are never expected to occur as keywords in any query. In addition, a relevant webpage may contain images, embedded audio, incoming and out-going hyper-links and meta-data, such as the title, the author, date of creation, perhaps even XML tags encoding its semantics. It certainly does not seem plausible that a rich representation like this could follow the same underlying distribution as a two-word query. If nothing else, the space that embodies the documents appears to be different than the space containing the queries.

3.1.1 Representation of documents and requests

To circumvent the above problem we are going to take a somewhat radical step – we are going to assume that queries and documents originate in an abstract space which is rich enough to represent the attributes of either. We will refer to this space as the **latent representation space**. A query, in its latent, unobserved form, contains all the attributes that a document might contain. The nature of these attributes depends on the specific collection we are dealing with, and on what aspects of the data we are trying to capture with our model. Let us consider a few examples. If we are dealing with a plain-text collection, the latent form of the query would include a complete, coherent narration concerning the subject of the underlying information need. If we are working with images, the latent query would be represented as a complete bitmap, along with elaborate textual description of what is pictured in that bitmap. Note that in this case, we assume that the documents (images) also contain this textual description, even if in reality we are dealing with a collection of images that are not annotated in any way. Similarly, in a video archive, both documents and queries contain a sequence of bitmap frames, the audio signal, augmented with a complete textual transcript of speech in the audio and a narration, describing the objects and actions in the frames. As another example, consider a question answering scenario: in this case we assume that our representation space consists of pairs that combine a given question with a correct answer. This, again, is a latent, unobserved representation of the data.

At some point before they are presented to us, both documents and queries undergo a certain deterministic transformation that converts them to the form that we actually observe. For example, in the case of a text-only collection, the documents remain unchanged, but the query transformation cuts out all the syntactic glue, removes duplicates and discards all but two or three salient keywords. If we are dealing with a collection of images, the transformation would apply to both documents and queries. Documents would be stripped of their textual description, leaving only the bitmap form. For queries, the transform would drop the image portion and reduce the description to a short list of keywords, as we described above. Similarly, in a video collection, documents would keep the frames and the audio signal, queries would keep some words from the transcript or the textual narrative. Finally, in a question-answering scenario questions would be stripped of their answers to form queries, and answers alone would become documents.

3.1.2 Advantages of a common representation

We admit, the above description assumes a fairly convoluted process for a simple fact of a few keywords making their way into the search box. On the other hand, we gain several distinct advantages by hypothesizing the process described above. These are:

1. **We can define a common generative model.** By assuming that documents and queries originate in the same space, we pave the way for defining a single distribution that can describe both documents and queries. This is an absolute necessity if we want to entertain the generative hypothesis.
2. **Anything can be used as a query in the model.** The fact that the query has an artificially enriched latent representation means that any aspect of that representation can initiate the searching process, as long as it happens to be observable. For example, we talked about image queries originating as a bitmap and narrative and then being cut down to a couple of keywords. But if the user happens to provide a sketch of what he is looking for, or perhaps a fragment of a similar image, that information can be used just as well in place of keywords. In fact, a fragment can

be used alongside the words if the user happens to provide both. After all, both are just two pieces of the same incomplete puzzle: the latent representation of the query.

3. **The model has natural extensions to cross-language, multimedia and semi-structured search.** It is almost self-evident that the proposed representation makes it very easy to model such tasks as cross-language or multi-language retrieval, multi-media retrieval or retrieval from semi-structured databases. For example, consider the cross-language retrieval problem, where documents are written in language A and queries issued in language B. In our approach, the latent representation of both documents and queries is a parallel narrative in both languages. Then, during the transformation to their observable form, documents would keep the narrative A, while queries would be reduced to a few keywords from the narrative B. For multi-lingual IR, where several languages are involved, the latent representation of each document/query includes a complete narrative in all languages involved. For semi-structured retrieval, each representation contains all the fields that are defined in the database schema, and, naturally, any subset of fields can be taken as observable and used as a query. In this respect, our model is markedly different from previous approaches to these problems. In the past research, most authors had to define completely new retrieval models to handle cross-language or multi-media scenarios. Two notable exceptions to this rule are the *inference network model* proposed by Turtle and Croft [135, 134], and the *translation model* of Berger and Lafferty [11]. Both models provide for graceful extensions to cross-language retrieval and allow some multi-media processing.
4. **The approach is robust for incomplete representations.** An important feature at the core of our approach is a single generative model underlying both the documents and the queries. As a natural side effect, we will be forced to construct a joint distribution over all forms of media present in a given document, or over all fields in a schema, if we are dealing with partially structured documents. This side effect becomes particularly important in semi-structured databases with missing or garbled information. In these cases having a joint generative model is a great advantage, since it will allow us to recover or fill in the missing fields based on the fields that were available. This recovery is very natural in the model: we simply assume that lost or garbled data is part of the transform that takes documents and queries from their latent space to the observable representation. We are not aware of many other models of semi-structured retrieval that provide for a similar recovery mechanism.
5. **Relevance feedback is a natural part of the model.** In a typical retrieval situation the only piece of information available to the system is the user's request, which represents a single sample from the relevant population. However, if it happens that the user is willing to provide another formulation of his request, or perhaps an example relevant document, the

generative hypothesis will allow us to handle the situation gracefully. Recall that the whole idea behind the GRH is that both user's requests and relevant documents are random samples from the same underlying model. Getting more examples of either relevant documents or requests will allow us to estimate the model with greater accuracy. However, we want to stress that requests may be used for feedback only if they reflect the current information need. This is slightly different from the query relevance feedback approach explored by Nallapati, Croft and Allan [96], where relevance of a given document to multiple queries was used to adjust that document's language model. In their case, different queries did not have to represent expressions of the same information need since they were estimating a model for that document only, as opposed to the *relevance* model for a particular information need.

6. **We do not have to factor the probability space.** To clarify this point, recall the classical probabilistic model of Robertson and Sparck Jones (section 2.3.2). User's query was not a formal part of the model, it was introduced only as a set of assumptions (section 2.3.2) to help estimate probabilities without relevant examples. Why was the query not included in the model? It turns out that formally introducing the query as a variable is not as simple as it may seem. Robertson and colleagues made a number of attempts, the best known is perhaps the *unified model* described in [115, 116]; for a more recent attempt see [14]. In every case the authors instantiate two distinct spaces: the space $\mathcal{Q}$ for queries and the space $\mathcal{D}$ for documents. A probabilistic model is then defined over the product space $\mathcal{Q}\times\mathcal{D}$. This seems fine on the surface, but in reality poses a very serious problem: queries and documents live in orthogonal dimensions of the space. The event that word "Washington" appears in the query has nothing in common with the event that the same word is frequent in some document. The former event concerns the $\mathcal{Q}$ part of the space, the latter is part of $\mathcal{D}$. There is no built-in notion of word overlap; it either has to be learned by observing relevant document-query pairs, or else we have to forcibly introduce *links* between dimensions in $\mathcal{D}$ and dimensions in $\mathcal{Q}$. Introducing these links is also not always a simple matter: Robertson [108, 109] discovered that working in the factored space can lead to certain consistency problems. This prompted him to argue for the necessity of imposing some sort of structure on top of $\mathcal{Q}\times\mathcal{D}$. Without delving further into the details, we would like to stress that the problems described above are the direct result of using the Cartesian product $\mathcal{Q}\times\mathcal{D}$ (factored space). The problems do not exist if we represent queries and documents in the same space. In our model, a word occurring in the query is the same type of event as the word occurring in a document.

3.1.3 Information retrieval under the generative hypothesis

In the above arguments we have suggested that we can view documents and queries as items that originate in the same representation space. This observation paves the way for translating the generative relevance hypothesis into an operational model of information retrieval. In this section we will discuss two options for constructing a retrieval model around the GRH. The first option involves invoking the GRH to approximate a probability distribution associated with the relevant class, which can be used directly with Robertson's probability ranking princinple. The second option uses the GRH to test a hypothesis of relevance for each document in the collection.

Using the probability ranking principle

A straightforward way to use GRH in constructing a retrieval system would be to integrate it with the classical probabilistic model of Robertson and Sparck Jones [117]. The model computes the probability ratio $\frac{P(\mathbf{d}|R=1)}{P(\mathbf{d}|R=0)}$ as a measure of relevance for each document $\mathbf{d}$ in the collection. In sections 2.3.2 and 2.3.2 we established that the primary weakness of the model arises from the need to compute $P(\mathbf{d}|R=1)$, the distribution associated with the relevant class, when no examples of relevant documents are available. In this section we will provide an intuition for how we can use GRH to estimate $P(\mathbf{d}|R=1)$.

We will start by accepting the generative hypothesis and assuming that all relevant documents are samples from the same underlying population as the query. Of course we know nothing about this population, or about the generative process responsible for generating it: the user provides us neither with examples of relevant documents nor with alternative re-formulations of the same query. However, we do have the user's original query, and according to the generative hypothesis this query $\mathbf{q}$ is a representative sample, drawn from the population of relevant items. We can use this fact to get some idea of what the relevant population might look like. Specifically, we will assume that the relevant population was sampled from a *relevance model*, a generative process, which can be expressed by a certain (quite general) formula with some unknown parameters θ. To estimate these parameters we will need a sample from the relevance model, and the user's query $\mathbf{q}$ will play the role of that sample. Once we have an estimate of θ_{rel} associated with the relevant class, we can directly use the relevance model $P(\mathbf{d}|\theta_{rel})$ in place of $P(\mathbf{d}|R=1)$ in the classical probabilistic model of Robertson and Sparck Jones (section 2.3.2). If our estimated relevance model indeed reflects the distribution of relevant documents, then we would achieve optimal ranking by ordering the documents according to the ratio $\frac{P(\mathbf{d}|R=1)}{P(\mathbf{d}|R=0)}$. We have to find a way to compute the denominator, but that should be fairly straightforward: as we argued before, almost all documents in the collection are non-relevant, so we can use collection-wide statistics to estimate non-relevant probabilities $P(\mathbf{d}|R=0)$. We have tested this form of ranking, and found that it does work reasonably well. However, we will argue later in this chapter that using the probability ratio $\frac{P(\mathbf{d}|R=1)}{P(\mathbf{d}|R=0)}$ for ranking may not always be optimal. This sub-optimality does not imply a flaw in the probability ranking principle itself; rather, it is a consequence of independence assumptions commonly made about the words in the document **d**. We will demonstrate that under these assumptions a better form of ranking may be derived from treating the retrieval process as a series of statistical hypothesis tests.

Retrieval through hypothesis testing

In this section we will outline a retrieval model that *directly* tests the generative relevance hypothesis for each document in the collection. Recall that the GRH postulates that relevant documents are random samples, drawn from the same population as the query. So if we wanted to determine whether a given document **d** is relevant to the user's query **q**, we would test how probable it is that **d** and **q** are samples from the same underlying population. Of course, as we discussed in the previous section, documents and queries may look very dissimilar in their observed form. This is where the common representation saves the day: we will view **q** and **d** as partial reflections of their complete latent representations. Since document and query representations share the same space, we can define a probability measure P over populations in that space and test whether **d** and **q** come from the same population. As with any statistical significance test, we will be deciding between two competing hypotheses:

H_0 : Our **null** hypothesis is that the document **d** and the query **q** were drawn from different, unrelated populations in the representation space.

H_R : Our **relevant** hypothesis is that **d** and **q** were drawn from the same population.

If our test indicates that H_R is significantly more probable that H_0, the GRH would lead us to believe that the document is in fact relevant to the query. If we could not reject H_0, we would be left to conclude that **d** is non-relevant.

This procedure can be extended to *ranking* an entire collection of documents in response to the query, as well as to selecting the most relevant *subset* of the collection. If ranked retrieval was our objective, we would present the documents in the order of decreasing probability of rejecting H_0. In other words, the documents at the top of the ranked list should be the ones that are most likely to have been sampled from the same population as the query. Under the GRH, this ranking criterion is equivalent to the probability ranking principle of Robertson[114], since GRH equates the event of relevance and the event of **d** and **q** being drawn from the same population.

It is also worthwhile to highlight the applicability of hypothesis testing to the problem of selecting *subsets* of documents to satisfy the query. For a query **q** and a given subset $\mathbf{d}_1, \ldots, \mathbf{d}_k$ of the collection, we define the following extensions of conjectures H_0 and H_R:

H'_0 : $\mathbf{q}, \mathbf{d}_1, \ldots, \mathbf{d}_k$ were drawn from unrelated populations
H'_R : $\mathbf{q}, \mathbf{d}_1, \ldots, \mathbf{d}_k$ were drawn from the same population

The set of documents for which we are most likely to reject H'_0 in favor of H'_R is the set most relevant to the query under the GRH. A detailed discussion of this procedure is beyond the scope of our work, but it would be instrumental if we wanted to extend our generative model to non-independent views of relevance discussed in section 2.1.5.

3.2 Formal Specification of the Model

We will now formally define the ideas presented in the previous section. We will start with the most general case and then outline specific representations for the various retrieval scenarios. Before we begin, we would like to stress that our formalism assumes a closed universe: there is a single user, performing a single task with a single underlying information need. Multiple queries may be involved, but they are all expressions of the same underlying need. Relevance is defined with respect to that information need only. Table 3.1 serves as a reference for symbols and notations used in this section.

Symbol	Meaning
$\mathcal{S} = \mathcal{S}_1 \times \mathcal{S}_2 \times \ldots \mathcal{S}_M$	latent representation space
$X = X_1 X_2 \ldots X_M$	a random variable representing an item in $\mathcal{S}$
$\mathbf{d} \in \mathcal{D}$	document $\mathbf{d}$ is a point in the document space $\mathcal{D}$
$\mathbf{q} \in \mathcal{Q}$	user's query $\mathbf{q}$ is a point in the query space $\mathcal{Q}$
$\mathbf{r}$	a relevant example, could be a document or a query
$D : \mathcal{S} \mapsto \mathcal{D}$	document generating function
$Q : \mathcal{S} \mapsto \mathcal{Q}$	query generating function
$R : \mathcal{D} \mapsto \{0, 1\}$	relevance judgment function
$\mathcal{C} \subset \mathcal{D}$	collection of observed documents
$\mathcal{D}_R = \{\mathbf{d} : R(\mathbf{d}) = 1\}$	population of relevant documents
$\mathcal{S}_R = D^{-1}(\mathcal{D}_R)$	population of relevant representations
$\mathcal{C}_R = \mathcal{D}_R \cap \mathcal{C}$	relevant documents observed in the collection
$P(\cdot)$	probability distribution over $\mathcal{S}$
Θ	de Finetti parameter space of P
$p(\theta)$	a probability density function over Θ
$P_{i,\theta}$	component distribution over dimension $\mathcal{S}_i$
$RM_{\mathbf{q}}$	relevance model estimated from the user's query $\mathbf{q}$
$RM_{\mathbf{d}}$	relevance model estimated from the document $\mathbf{d}$
$M_{\mathbf{d}}$	distribution based on word frequencies in $\mathbf{d}$

Table 3.1. Notation: frequently used symbols and expressions.

3.3 Representation of Documents and Queries

We assume that in their unobserved form both documents and queries can be represented as a sequence of M random variables $X_1 \ldots X_M$. Each X_i takes values in some appropriate space $\mathcal{S}_i$, so the latent representation space $\mathcal{S}$ we discussed in the previous section takes the form of a Cartesian product $\mathcal{S}_1 \times \ldots \times \mathcal{S}_M$. In preparation for what follows, it may help the reader to think of $\mathcal{S}$ as the space that contains the most complete representation possible for any given document or query – a kind of information space, or knowledge space. This, of course, is only an analogy; we make no attempt to represent meaning.

3.3.1 Document and query generators

We define a *document generating transform* D to be a function that converts a complete representation $\mathbf{x} = x_1 \ldots x_M$ to a form in which a document $\mathbf{d}$ might actually be observed in the collection; formally $D : \mathcal{S} \rightarrow \mathcal{D}$. Similarly, we define the function $Q : \mathcal{S} \rightarrow \mathcal{Q}$ to be the *query generating transform* – a mapping from representations in the complete space $\mathcal{S}$ to something that looks like a user query $\mathbf{q}$. We would like to stress that there is no randomness involved in either $D(\mathbf{x})$ or $Q(\mathbf{x})$ – they are normal deterministic functions like $\log(x)$ or $\sqrt{x}$. And for reasons to be detailed later we will want these functions to take a particularly simple form. The main effect of D and Q will be that they take the sequence $x_1 \ldots x_M$ and remove some of the variables from that sequence. The result of either D or Q will be a new shorter sequence $x_{i_1} \ldots x_{i_m}$ that contains some of the elements of the original representation. In a sense, D and Q discard parts of the rich representation that are not supposed to appear in documents and queries respectively. For example, D may strip images of their captions, and Q may remove all but a handful of keywords from a long detailed narration. As a consequence, we observe that the document space $\mathcal{D}$ and the query space $\mathcal{Q}$ are both contained in the full representation space. To get from $\mathcal{S}$ to $\mathcal{Q}$ all we have to do is lose a few dimensions. Note that spaces $\mathcal{D}$ and $\mathcal{Q}$ may or may not overlap.

3.3.2 Relevant documents

By a *collection* $\mathcal{C}$ we will mean a finite set of points in the document space $\mathcal{D}$ – the points that correspond to documents that we actually have in our database. We define *relevance judgments* to be the function $R : \mathcal{D} \rightarrow \{0,1\}$ that specifies whether a given document $\mathbf{d}$ is going to be judged relevant or non-relevant. We assume that R is deterministic, but cannot be observed unless the user provides us with relevant examples. The relevance function R involves only documents, a query is unnecessary since we are operating in a closed universe with a single underlying information need. The space of all documents that would be judged relevant if they existed is $\mathcal{D}_R = \{\mathbf{d} \in \mathcal{D} :$

$R(\mathbf{d})=1\}$. Relevant documents that actually exist in our collection form the *relevant set* $\mathcal{C}_R = \mathcal{D}_R \cap \mathcal{C}$.

3.3.3 Relevance in the information space

A reader particularly interested in the nature of relevance may wonder why we chose to define relevance over the space $\mathcal{D}$ of observable documents, when in fact we have a much more general *information* space $\mathcal{S}$. Any assessor would certainly find it easier to judge relevance based on the complete, uncut representations, say an image with a detailed narrative, as opposed to just a bitmap. However, we want our model to reflect reality whenever possible, and in reality a relevance assessor only has access to the observable space $\mathcal{D}$. Furthermore, defining the relevance function over $\mathcal{S}$ may lead to cases where we are unable to decide whether an observed document $\mathbf{d}\in\mathcal{D}$ is relevant or not. Such cases may arise because the document generating function D is not a one-to-one mapping, and a given $\mathbf{d}$ may have several pre-images $D^{-1}(\mathbf{d})$, some relevant, some not. While this ambiguity may sound attractive for modeling the natural uncertainty of relevance judgments, it would require a cumbersome tying of the functions R and D, which we prefer to avoid. In addition, we would like to stress that a relevance function R defined over the document space automatically induces a relevance function over the representation space: a composition $R(D(\mathbf{x}))$ is a well-defined function for every representation $\mathbf{x}\in\mathcal{S}$. The relevant population in the information space is defined as $\mathcal{S}_R = D^{-1}(\mathcal{D}_R)$. Intuitively, it consists of all latent representations $\mathbf{x}\in\mathcal{S}$ that would be judged relevant once they got transformed into documents $\mathbf{d} = D(\mathbf{x})$.

3.3.4 Relevant queries

Now that we have defined a notion of relevance on the information space $\mathcal{S}$, we can talk about what it means to be a *relevant request*; i.e. a request expressing the underlying information need. Getting a sample of what a relevant request *might* look like is actually very straightforward in our model. We already know what portion of the information space would be judged relevant ($\mathcal{S}_R = D^{-1}(\mathcal{D}_R)$), so we can simply apply the query generating function $Q(\mathbf{x})$ to every point $\mathbf{x}$ in $\mathcal{S}_R$ to get the corresponding set $\mathcal{Q}_R$ of all queries that reflect the information need. Accordingly, if we wanted to know if some request $\mathbf{q}$ was relevant or not, we could look at the inverse $Q^{-1}(\mathbf{q})$ and check to see if it falls into $\mathcal{S}_R$ or not. However, we might ask whether $Q^{-1}(\mathbf{q})$ can really tell us anything about relevance. The answer is probably no, and the reason hides in the fact that the function Q mapping representations to queries is not a one-to-one mapping. By its very nature Q is a many-to-one function: as we discussed above, the effect of Q is to discard some dimensions of the information space, so all representations $\mathbf{x}$ that differ on the discarded dimensions will be collapsed into the same query $\mathbf{q} = Q(\mathbf{x})$. For example, if the effect of Q was to retain only the first word in a narrative, all possible texts starting with "the" would be collapsed to the same query. This means that the inverse Q^{-1} is not a function, it is a one-to-many mapping. For any given request $\mathbf{q}$ there will be many pre-images $Q^{-1}(\mathbf{q})$ in the information space, some of them might fall into $\mathcal{S}_R$, most probably will not. It is not possible to definitively say whether any given request is relevant. Fortunately, this inability is not at odds with reality: any human assessor would be hard-pressed to state whether or not a typical query definitively represents the underlying information need. He may say that a request is likely or not likely to express his need; in our model this likelihood would be reflected by the amount of overlap between $Q^{-1}(\mathbf{q})$ and the relevant sub-space $\mathcal{S}_R$.

3.3.5 Summary of representations

Before we dive into the probabilistic aspects of relevance, we would like to provide a brief summary of the representation choices we made in our model. An informal diagram of these choices is provided in Figure 3.1. We made the following assumptions:

1. Documents and queries originate in the same latent representation space $\mathcal{S}$. We can think of $\mathcal{S}$ as *information* or *knowledge* space; it is rich enough to encode all desired attributes of documents and queries (including non-existent attributes).
2. A query is the result of applying a function Q to some point $\mathbf{x}$ in the information space. $Q(\mathbf{x})$ filters and shuffles components of $\mathbf{x}$ to make it look like a query. To get a document, we apply a different transform $D(\mathbf{x})$. Both D and Q are deterministic.
3. A user's information need corresponds to a *relevant population* $\mathcal{S}_R$. Relevant documents and requests result from applying D and Q to points in that population.

3.4 Probability Measures

In the previous section we presented a formalism for embedding documents and queries in a common representation space. We will now introduce a probability distribution over this space. The purpose of the distribution is to describe how likely we are to observe a given document $\mathbf{d}$ or a query $\mathbf{q}$. In our case, both documents and queries are deterministic transformations of the information representations $\mathbf{x}\in\mathcal{S}$. This fact makes it natural to define the generative model on the space of complete representations, and then extend it to documents and queries using the functions D and Q. We stress that the distribution will be defined over the entire space, not just over the relevant population. Distributions over relevance-based events will be addressed in the following section.

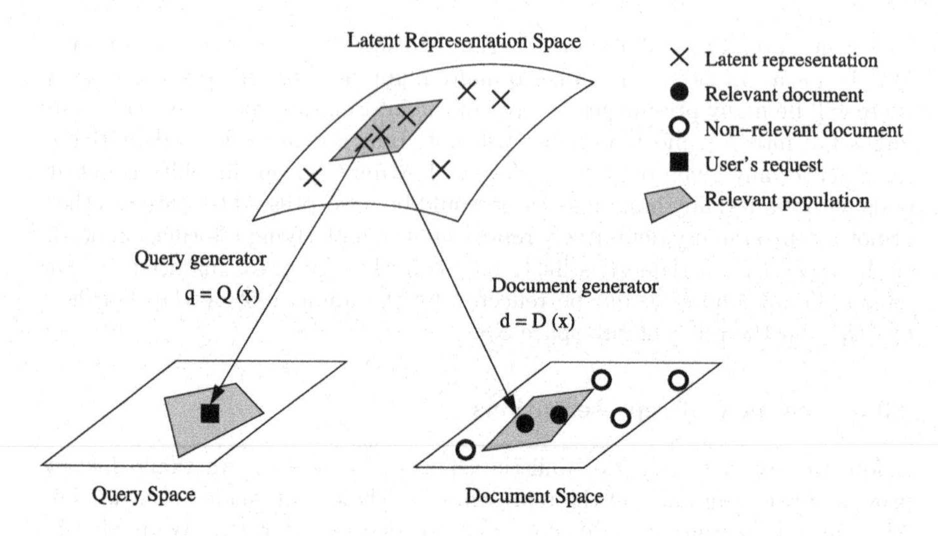

Fig. 3.1. Representation of documents, queries and relevance in our model. We assume a latent representation space which contains a relevant population. All documents and queries are deterministic transformations of points in that latent space. Relevant documents and queries are samples drawn from the relevant population.

3.4.1 Distribution over the representation space

Let $P : \mathcal{S} \mapsto [0, 1]$ denote the joint probability distribution over the variables $X_1 \ldots X_M$. In order to proceed, we have to make a certain simplifying assumption about the random variables $X_1 \ldots X_M$ that make up our information space. The first thing that comes to mind is to follow the classical probabilistic model and assume that X_i are mutually independent. However, independence is a rather strong assumption, and it will needlessly restrict our model. Instead, we are going to assume that $X_1 \ldots X_M$ are *exchangeable* or *order-invariant*. To illustrate the concept of exchangeability, let us suppose for a moment that all X_i take values in the same space $\mathcal{S}_1 = \mathcal{S}_2 = \ldots = \mathcal{S}_M$. Informally this might mean that all X_i are words from the same vocabulary, or that all X_i are numbers in the same range. When we say that X_i are *exchangeable*, we mean that the *order* of observations does not matter. For example, if X_1, X_2, X_3 represent the first, second and third words in a phrase, assuming exchangeability means that the phrase "New York Yankees" will have the same probability as "York Yankees New", and so will any other re-ordering of these words. Exchangeability of the variables does not in any way imply their independence.

We can have very strong dependencies between the words[1], as long as these dependencies are not affected by the order in which the words appear.

If we want to be absolutely precise in our definitions, we are dealing with *partial* exchangeability, which means that we can arbitrarily re-order some of the variables, but not all of them. We will be concerned with a very restricted form of partial exchangeability, necessitated by the fact that some variables may take values that are incompatible with other variables. For example, suppose we are dealing with an image database where $X_1 \ldots X_{k-1}$ represent some visual features of the image, and $X_k \ldots X_n$ represent the textual description. In this case we cannot allow any ordering that assigns an image feature $(x_{i<k})$ to a word variable $(X_{j \geq k})$. At the same time, we do allow any re-ordering within $X_1 \ldots X_{k-1}$, and within $X_k \ldots X_n$. For all practical purposes, this restricted form of partial exchangeability is no different from complete exchangeability; all major theoretical results hold without modification.

According to de Finetti's representation theorem, if $X_1 \ldots X_M$ are exchangeable, their joint distribution can be expressed in the following form:

$$P(X_1{=}x_1 \ldots X_M{=}x_m) = \int_\Theta \left\{ \prod_{i=1}^{M} P_{i,\theta}(x_i) \right\} p(\theta) \mathrm{d}\theta \tag{3.1}$$

The variable θ in equation (3.1) is a vector of hidden parameters. If we knew the exact value of θ, the variables X_i would be mutually independent, hence the joint probability decomposes into a product of the marginals $P_{i,\theta}(x_i)$. But θ is unknown, so we integrate over the space of all possible parameter settings Θ. The quantity $p(\theta)$ is a probability measure that tells us which parameter settings are a-priori more likely to produce a document or a query. Each $P_{i,\theta}(x_i)$ is an appropriate probability distribution over one dimension $\mathcal{S}_i$ of our representation space $\mathcal{S}$. Note that if all dimensions $\mathcal{S}_i$ are the same (all words, or all numbers), then X_i are also identically distributed given θ.

3.4.2 Distribution over documents and queries

In the previous section we defined a probability distribution for representations in the information space $\mathcal{S}$. In this section we will show how the distribution can be extended to observable documents and queries. As we discussed in section 3.3.1, documents and queries are deterministic transformations of information representations $\mathbf{x} \in \mathcal{S}$. So we can imagine the following two-step process for generating a given document $\mathbf{d}$: first, we pick a representation

[1] Consider the famous *Polya's urn*, also known as the *contagion model.* We have an urn with two words "a" and "b". We pull out a word, duplicate it and put both duplicates back into the urn. The result of the second draw is certainly not independent of what word we pull out on the first draw: if we pull out "a" than pulling it out again is twice as likely as pulling out "b". But the joint probability of a given sequence of observations will not depend on the order in which they appear: $P(a, a, b) = \frac{1}{2}\frac{2}{3}\frac{1}{4}$ is the same as $P(b, a, a) = \frac{1}{2}\frac{1}{3}\frac{2}{4}$, etc.

$\mathbf{x}\in\mathcal{S}$ with probability $P(\mathbf{x})$; then we apply the document generating transform to get the observation $\mathbf{d}=D(\mathbf{x})$. However, since D is not a one-to-one function, there may be many different representations $\mathbf{x}$ mapping to the same document $\mathbf{d}$ (think of all possible textual descriptions that could go together with a given image). So to get the overall probability of getting a particular document $\mathbf{d}$, we need to take into account all representations $\mathbf{x}$ that could be collapsed into $\mathbf{d}$:

$$\begin{aligned} P(D(X)=\mathbf{d}) &= P(X\in D^{-1}(\mathbf{d})) \\ &= \sum_{\mathbf{x}:D(\mathbf{x})=\mathbf{d}} P(X=\mathbf{x}) \end{aligned} \tag{3.2}$$

We will apply the same generative process to explain where queries come from. To get a query $\mathbf{q}$ we first sample a representation $\mathbf{x}\in\mathcal{S}$ with probability $P(\mathbf{x})$ and then convert it to a query by setting $\mathbf{q}=Q(\mathbf{x})$. And as was the case with documents, the overall probability of getting $\mathbf{q}$ depends on all points $\mathbf{x}$ that would be mapped to $\mathbf{q}$, so:

$$\begin{aligned} P(Q(X)=\mathbf{q}) &= P(X\in Q^{-1}(\mathbf{q})) \\ &= \sum_{\mathbf{x}:Q(\mathbf{x})=\mathbf{q}} P(X=\mathbf{x}) \end{aligned} \tag{3.3}$$

In general, if Q and D were arbitrary functions, the formulas for $P(\mathbf{d})$ and $P(\mathbf{q})$ could end up being quite complex, and in fact might look very different from each other. But recall that in section 3.3.1 we restricted D and Q to a particularly simple form. The document transform D takes a latent representation $\mathbf{x} = x_1x_2\ldots x_M$ and strips out the components that are not supposed to appear in documents. The result is a sub-sequence $x_{i_1}x_{i_2}\ldots x_{i_n}$, where $n < M$. The query transform Q performs the same type of operation resulting in a sub-sequence $x_{j_1}x_{j_2}\ldots x_{j_m}$, $m < M$. Below we are going to show that restricting D and Q in this fashion leads to a very important equivalence: the generative model of documents $P(\mathbf{d})$ will take the same functional form as the generative model of queries $P(\mathbf{q})$. Furthermore, the functional form will be the same as the distribution over information representations $P(\mathbf{x})$. Below is the derivation concerning the documents. Suppose D keeps components $i_1\ldots i_n$ from the initial representation and removes components $i_{n+1}\ldots i_M$. Then the probability of observing a document $\mathbf{d} = x_{i_1}x_{i_2}\ldots x_{i_n}$ is:

$$\begin{aligned} P(\mathbf{d}) &= \sum_{\mathbf{x}:D(\mathbf{x})=\mathbf{d}} P(X=\mathbf{x}) \\ &= \sum_{\mathbf{x}\in\mathcal{S}} \delta(\mathbf{d}, D(\mathbf{x})) \cdot P(X=\mathbf{x}) \\ &= \int_{\Theta} \left\{ \sum_{\mathbf{x}\in\mathcal{S}} \delta(\mathbf{d}, D(\mathbf{x})) \prod_{i=1}^{M} P_{i,\theta}(x_i) \right\} p(\theta)d\theta \end{aligned}$$

$$= \int_{\Theta} \left\{ \sum_{x_1} \cdot\cdot \sum_{x_M} \underbrace{\delta(d_1, x_{i_1}) \cdots \delta(d_n, x_{i_n})}_{\delta(\mathbf{d}, D(\mathbf{x}))} \cdot P_{1,\theta}(x_1) \cdots P_{M,\theta}(x_M) \right\} p(\theta) \mathrm{d}\theta$$

$$= \int_{\Theta} \underbrace{\prod_{a=1}^{n} \left(\sum_{x_{i_a}} \delta(d_a, x_{i_a}) P_{i_a,\theta}(x_{i_a}) \right)}_{\text{dimensions retained by } D} \underbrace{\prod_{b=n+1}^{M} \left(\sum_{x_{i_b}} P_{i_b,\theta}(x_{i_b}) \right)}_{\text{discarded by } D} p(\theta) \mathrm{d}\theta$$

$$= \int_{\Theta} \left\{ \prod_{a=1}^{n} P_{i_a,\theta}(d_a) \right\} p(\theta) \mathrm{d}\theta \tag{3.4}$$

The first step of the derivation comes from the definition of $P(D{=}\mathbf{d})$ in equation (3.2). The only change we made was to introduce the indicator function $\delta(\mathbf{d}, D(\mathbf{x}))$ which equals one when $\mathbf{d}{=}D(\mathbf{x})$ and zero for any $\mathbf{x}$ which does not map to $\mathbf{d}$. This gives the same effect as the subscript $\mathbf{x} : D(\mathbf{x}){=}\mathbf{d}$ in equation (3.2). In the second step, we substitute the definition of the relevance model P from equation (3.1). We use the Fubini-Tonelli theorem to move the integral $\int_{\Theta}$ outside the summation.[2] The third step in the derivation is the result of expanding the information representation $\mathbf{x}$ into its full form $x_1 \ldots x_M$. The summation over all $\mathbf{x}{\in}\mathcal{S}$ becomes a nested summation over all dimensions of $\mathcal{S}$. We also unravel the indicator function $\delta(\mathbf{d}, D(\mathbf{x}))$ to show that d_a must be equal to x_{i_a} for all document components $a{=}1 \ldots n$. Recall that x_{i_a} is the a'th element in the sub-sequence that results from applying D to the information representation $x_1 \ldots x_M$. In the fourth step of equation (3.4) we move the components $\delta(d_a, x_{i_a})$ and $P_{i_a,\theta}(x_{i_a})$ towards their respective summations $\sum_{x_{i_a}}$. We can do this because these terms only depend on the values x_{i_a}. Then we note that all summations are in effect disjoint, and we can therefore group them into two products: the first involving the dimensions $\mathcal{S}_i$ that are allowed to occur in the documents and the second containing the dimensions that are discarded by D. To get the final step, we observe that the second product involves the summations of $P_{i_b,\theta}(x_{i_b})$ over all possible values of x_{i_b}. Since every $P_{i_b,\theta}$ is a probability distribution, all summations will add up to one, and the product of $(M{-}n)$ ones equals one. Finally, the indicator function $\delta(d_a, x_{i_a})$ will reduce every summation in the first product to a single term, specifically the term $x_{i_a} = d_a$. This gives us the final form of equation (3.4). If we assume that the query transform Q operates in the same way as D and reduces a representation $x = x_1 \ldots x_M$ to a sub-sequence $x_{j_1} \ldots x_{j_m}$, then we can carry out a similar argument to demonstrate that the query distribution takes the form:

[2] The Fubini-Tonelli theorem is necessary for exchanging the integral and the summation if the representation space $\mathcal{S}$ is not a finite set. The theorem holds as long as $P_{i,\theta}()$ and $p(\theta)$ are "nice" (measurable) functions.

$$P(Q=\mathbf{q}) = \sum_{\mathbf{x} \in \mathcal{S}} \delta(\mathbf{q}, Q(\mathbf{x})) \cdot P(X=\mathbf{x})$$
$$\dots$$
$$= \int_{\Theta} \left\{ \prod_{b=1}^{m} P_{j_b,\theta}(q_b) \right\} p(\theta) d\theta \qquad (3.5)$$

An important thing to observe is that equation (3.5) has the same form as equation (3.4), and, for that matter, equation (3.1). The only difference is that the product in each case goes over a different set of component dimensions. So we can conclude that the generative model of documents $P(\mathbf{d})$ has the same general form as the generative model of queries $P(\mathbf{q})$, as long as documents and queries are projections from the same latent representation space.

3.4.3 Significance of our derivations

Why go through all the trouble to get the result stated above? Why is it important that $P(\mathbf{d})$ look like $P(\mathbf{q})$? The reason goes to the heart of the assumption made in the generative hypothesis. The hypothesis states that queries and relevant documents are samples from the same underlying model. Without a good justification, this claim will draw an uncomfortable amount of criticism from Information Retrieval practitioners. On the surface, the GRH seems to claim that documents and queries will be identically distributed, which is certainly a false assumption. Documents do not look like queries, they have different characteristics, and claiming that they are samples from the same model appears ludicrous. The sole purpose of our argument is to show that the generative hypothesis is not an absurdity. Documents and queries can and will look different under the GRH. In fact, their observable forms can be completely different: a document can be a bitmap and a query can be a string of text. However, as we demonstrated in the previous sections, we can still formally treat them as samples from the same generative model. We force documents and queries to originate in the same space: the "representation" space, which is general enough to represent both. Then we use the functions D and Q to transform the information representation in such a way that queries look like queries and documents look like documents. Very conveniently, these transformations do not change the functional form of the distribution. So, despite their different appearance and different characteristics, queries and relevant documents can be formally treated as samples from the same underlying probability distribution.

3.4.4 Summary of probability measures

Let us summarize the probability measures we developed in this section. We show an informal diagram of our model in Figure 3.2.

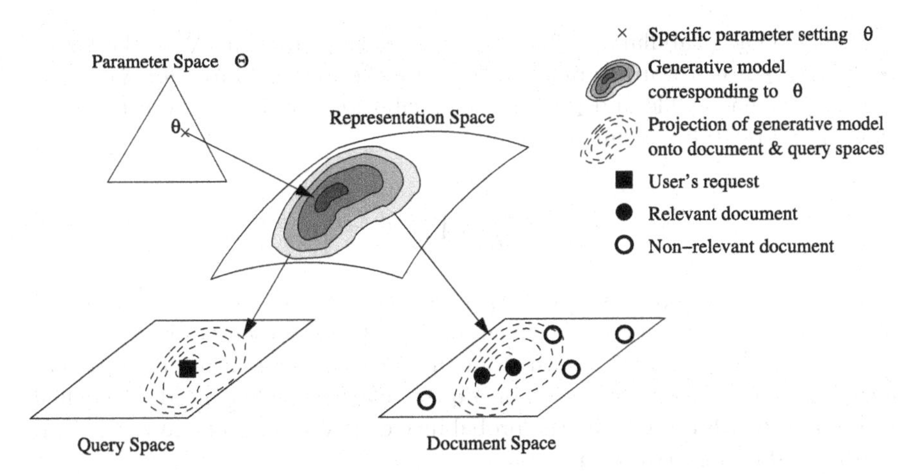

Fig. 3.2. A generative model is a probability distribution over latent representation space $\mathcal{S}$. Dimensions of $\mathcal{S}$ are exchangeable, so probabilities involve a hidden parameter θ. The generative process has natural projections to observable document and query spaces under transforms D and Q.

1. We defined a generative model P over the latent representation space. The purpose of P is to describe what document and query representations are likely to be observed during random sampling.
2. We assumed that random variables X_i in the latent representation are *exchangeable* or order-invariant. By de Finetti's representation theorem, the generative model P will depend on a hidden parameter θ and will take the form given by equation (3.1).
3. We discussed how the document and query transforms D and Q allow us to extend P from latent representations to observable documents and queries. We showed that for our choices of D and Q, generative models of queries and documents will have the same functional form. We used this to justify the generative relevance hypothesis.

3.5 Relevance Models

In this section we will provide a formal definition for the concept of a *relevance model* , which will be used extensively in the remainder of the book. Informally, the relevance model is our best approximation of the model that describes *relevant* documents and queries reflecting the user's information need. Formally, it is a point-estimate of the generative process for the relevant population: $P(\cdot|R{=}1)$. We will start by introducing relevance models in a restricted domain, where both documents and queries are strings in the same language. This restriction makes the initial definitions cleaner, and we will ultimately relax it towards the end of this section.

Let $\mathcal{V}$ be a vocabulary, and assume that each component X_i of the latent representation is a word from $\mathcal{V}$. In this case the random variables $X_1 \ldots X_M$ are *fully* exchangeable and their joint distribution, originally given by equation (3.1), now takes a simpler form:

$$P(X_1 \ldots X_M) = \int_\Theta \left\{ \prod_{i=1}^{M} P_\theta(X_i) \right\} p(\theta) d\theta \tag{3.6}$$

Note that we now have a single distribution $P_\theta(\cdot)$ for all components, instead of a separate $P_{i,\theta}(\cdot)$ for each dimension i of the representation space. Let $\mathbf{r} = r_1 \ldots r_m$ be a sample drawn from the *relevant* population, e.g. a document relevant to the user's need, or a query expressing that need. We define **relevance model** of $\mathbf{r}$ to be the probability distribution over the words in $\mathcal{V}$ conditioned on observation $\mathbf{r}$:

$$RM_{\mathbf{r}}(\cdot) = P(\cdot | r_1 \ldots r_m) = \frac{P(\cdot, r_1 \ldots r_m)}{P(r_1 \ldots r_m)} \tag{3.7}$$

The joint distributions in the numerator and denominator follow equation (3.6), which allows us to give a complete expression for the relevance model of $\mathbf{r}$:

$$RM_{\mathbf{r}}(\cdot) = \frac{\int_\Theta P_\theta(\cdot) \left\{ \prod_{i=1}^{m} P_\theta(r_i) \right\} p(\theta) d\theta}{\int_\Theta \left\{ \prod_{i=1}^{m} P_\theta(r_i) \right\} p(\theta) d\theta} \tag{3.8}$$

Our next goal is to connect the relevance model $RM_{\mathbf{r}}$ to the true probability distribution of documents in the relevant class, i.e. $P(\cdot | R{=}1)$. To aid the reader in making this connection, we will suggest two intuitive interpretations of equation (3.8). The first one, based on the ideas of co-occurrence, will appeal to a frequentist reader, while the second will have more of a Bayesian flavor.

3.5.1 Frequentist interpretation: a sampling game

We are trying to learn something about the relevant population. While we don't know the law of that population, but we are allowed to draw repeated samples from it. Suppose that we have drawn m single-word samples from the relevant population and observed the words $r_1 \ldots r_m$. If we make one more sample, what is the likelihood that we will observe the word v?

The best way to estimate the probability of seeing v on the next draw would be to look at how likely v is to *co-occur* with the words we have already observed. In other words, we look at the joint probability of observing v together with $r_1 \ldots r_m$. That is precisely the reasoning behind equation (3.7), and equation (3.6) provides us with a mechanism to compute that probability of co-occurrence. Furthermore, since $\mathbf{r}$ is the only available sample of words from the relevant population, $P(v|\mathbf{r})$ is our best estimate of $P(v|R{=}1)$, which is the true likelihood of seeing v during a random sample from the relevant population.

3.5.2 Bayesian interpretation: uncertainty about relevance

Let's take a detailed look at equation (3.6), which specifies the process responsible for generating all document and query representations. The process relies on an underlying vector of parameters θ: to get a representation $\mathbf{x}$ we first pick θ and then generate $\mathbf{x}$ from $P_\theta(\cdot)$. Different parameter settings θ will generate different sub-populations within the representation space, and it is natural to suppose that there exists a parameter setting θ_R that will generate the relevant sub-population $\mathcal{S}_R$. Of course, we don't know the magic value of θ_R. However, we do have a sample $\mathbf{r}$ drawn from the relevant population, and it stands to reason that $\mathbf{r}$ was generated by θ_R, along with all other points in $\mathcal{S}_R$. We can use $\mathbf{r}$ together with Bayes' rule to guess what values of θ could have generated $\mathbf{r}$, and hopefully this will give us some idea of what the elusive θ_R might look like. Specifically, we will construct a posterior distribution over θ, conditioned on having observed $\mathbf{r} = r_1 \ldots r_m$:

$$
\begin{aligned}
p(\theta|\mathbf{r}) &= \frac{P(r_1 \ldots r_m|\theta)p(\theta)}{P(r_1 \ldots r_m)} \\
&= \frac{\{\prod_{i=1}^{m} P_\theta(r_i)\}\, p(\theta)}{\int_\Theta \{\prod_{i=1}^{m} P_\theta(r_i)\}\, p(\theta) \mathrm{d}\theta} \qquad (3.9)
\end{aligned}
$$

It is illustrative to compare the posterior distribution $p(\theta|\mathbf{r})$ to the prior from equation (3.6). The prior $p(\theta)$ reflects the general uncertainty in selecting a set of parameters, prior to any observation. It should embed enough diversity to allow the generation of any population of documents/queries, including a plethora of non-relevant populations. On the other hand, the posterior $p(\theta|\mathbf{r})$ will be updated to reflect the relevant observation $\mathbf{r}$, and we expect it to assign higher probability to parameter settings similar to θ_R, which are more likely to generate other relevant representations. If we replace the prior $p(\theta)$ with the posterior $p(\theta|\mathbf{r})$ in equation (3.6), we will obtain a new generative process that assigns higher likelihood to relevant documents/queries and lower likelihood to non-relevant representations. This new process can be considered a model of the relevant population $\mathcal{S}_R$, and we will denote it as $P(\cdot|\mathbf{r})$. It is easy to verify that the word distribution induced by this new generative process is precisely the relevance model $RM_\mathbf{r}$:

$$
\begin{aligned}
P(\cdot|\mathbf{r}) &= \int_\Theta P_\theta(\cdot) p(\theta|\mathbf{r}) \mathrm{d}\theta \\
&= \int_\Theta P_\theta(\cdot) \frac{\{\prod_{i=1}^{m} P_\theta(r_i)\}\, p(\theta)}{\int_\Theta \{\prod_{i=1}^{m} P_\theta(r_i)\}\, p(\theta) \mathrm{d}\theta} \mathrm{d}\theta \\
&= RM_\mathbf{r}(\cdot) \qquad (3.10)
\end{aligned}
$$

3.5.3 Multi-modal domains

In the beginning of this section we expressly restricted our discussion to the fully-exchangeable domain, where every dimension $\mathcal{S}_i$ is represented by the same vocabulary $\mathcal{V}$. This restriction is undesirable because it limits our applications to single-language retrieval tasks. Fortunately, it is easy to extend the definition of relevance models to partially-exchangeable cases, including cross-language, structured and multimedia domains.

Let $\mathbf{r} = r_1 \ldots r_m$ be a sample from the relevant population, and suppose that dimensions of the representation space can be partitioned into sets $\mathbf{a}=\{1 \ldots k\}$ and $\mathbf{b}=\{k+1 \ldots m\}$. The two sets are intended to represent two distinct "modalities" of the data, such as words from two different languages, or two incompatible feature sets in a multimedia domain. Dimensions within each set are exchangeable, but we cannot assign a value r_b to a random variable X_a whenever $a \in \mathbf{a}$ and $b \in \mathbf{b}$. We define two vocabularies $\mathcal{V}_\mathbf{a}$ to provide values for dimensions in the set $\mathbf{a}$ and similarly $\mathcal{V}_\mathbf{b}$ for dimensions in the set $\mathbf{b}$. In this setting the joint probability of observation $r_1 \ldots r_m$ will be expressed as:

$$P(r_1 \ldots r_m) = \int_\Theta \left\{ \prod_{i \in \mathbf{a}} P_{\mathbf{a},\theta}(r_i) \right\} \left\{ \prod_{j \in \mathbf{b}} P_{\mathbf{b},\theta}(r_j) \right\} p(\theta) d\theta \tag{3.11}$$

and the posterior distribution over parameters θ, conditioned on $r_1 \ldots r_m$ will take the following form:

$$p(\theta | r_1 \ldots r_m) = \frac{\left\{ \prod_{i \in \mathbf{a}} P_{\mathbf{a},\theta}(r_i) \right\} \left\{ \prod_{j \in \mathbf{b}} P_{\mathbf{b},\theta}(r_j) \right\} p(\theta)}{\int_\Theta \left\{ \prod_{i \in \mathbf{a}} P_{\mathbf{a},\theta}(r_i) \right\} \left\{ \prod_{j \in \mathbf{b}} P_{\mathbf{b},\theta}(r_j) \right\} p(\theta) d\theta} \tag{3.12}$$

The only difference from the posterior distribution for the fully exchangeable case is than now we have two component distributions: $P_{\mathbf{a},\theta}$ and $P_{\mathbf{b},\theta}$, corresponding to the two modalities $\mathbf{a}$ and $\mathbf{b}$. It is important to realize that the relevant sample $\mathbf{r}$ does not necessarily have to have components from both $\mathbf{a}$ and $\mathbf{b}$: one of these sets is allowed to be empty. For example, in a cross-language retrieval scenario a relevant sample may only be available in one of the two languages involved. Using the posterior given in equation (3.12) we define two relevance models:

$$RM_\mathbf{r}^\mathbf{a}(\cdot) = \int_\Theta P_{\mathbf{a},\theta}(\cdot) p(\theta | r_1 \ldots r_m) d\theta \tag{3.13}$$

$$RM_\mathbf{r}^\mathbf{b}(\cdot) = \int_\Theta P_{\mathbf{b},\theta}(\cdot) p(\theta | r_1 \ldots r_m) d\theta \tag{3.14}$$

Note that a separate relevance model is required for each modality because their vocabularies $\mathcal{V}_\mathbf{a}$ and $\mathcal{V}_\mathbf{b}$ are not interchangeable. The arguments above can be easily extended to a case where $k > 2$ modalities are present in the data: equations (3.11, 3.12) will involve a separate product for each of the k modalities, and k separate relevance models will be constructed using equation (3.14).

3.5.4 Summary of relevance models

In this section we defined a *relevance model* to be the probability distribution associated with the relevant population. We described a procedure for estimating a relevance model from a sample $\mathbf{r}$, which could be a relevant document or a query expressing the user's information need. The estimate is based on a joint probability of observing a given word v together with the relevant sample $\mathbf{r}$. We provided two interpretations of relevance models – from the Bayesian and the frequentist perspective – and argued why our estimate can indeed be considered a model of the relevant population. Finally, we briefly outlined how relevance models can be estimated in partially-exchangeable domains, such as cross-language and multimedia retrieval settings: a separate relevance-based distribution is constructed for each modality (e.g. for each language).

3.6 Ranked Retrieval

In this section we will provide the last component necessary for translating our framework into an operational retrieval system: a criterion for ranking documents $\mathbf{d}$ in response to the user's query $\mathbf{q}$. We will discuss several ranking criteria, all of which will be based on the concept of relevance model, introduced in section 3.5. For the extent of this section we will limit our discussion to the mono-lingual case where both the query and the document are strings of words in the same language. Multi-modal extensions are straightforward but differ from one case to another, so we chose to postpone their discussion until chapter 5.

Recall that in section 3.1.3 we informally discussed two document ranking principles that could be constructed around the generative relevance hypothesis: one based on Robertson's probability ranking principle, the other based on hypothesis testing. We will develop ranking criteria around both of these ideas.

3.6.1 Probability ranking principle

Let us start by discussing the ranking criterion that was used in the classical probabilistic model of Robertson and Sparck Jones [117]. As we discussed in section 2.3.1, we can achieve optimal retrieval performance if we ranked the documents $\mathbf{d}$ according to the posterior probability of relevance $P(R{=}1|\mathbf{d})$. In section 2.3.2 we explained that the probability of relevance is rank-equivalent to the probability ratio $\frac{P(\mathbf{d}|R=1)}{P(\mathbf{d}|R=0)}$, so theoretically using this ratio should lead to the best retrieval results. In practice the quality of ranking heavily depends on how accurately we can estimate the probability in the numerator. Most previous publications (sections 2.3.2 and 2.3.2) used heuristic estimates of the probability $P(\mathbf{d}|R{=}1)$ when no relevant examples were available. We will now show that the generative hypothesis makes it possible to compute the estimates in a more formal fashion.

In the previous sections we defined the relevance model $RM_{\mathbf{r}}$ to be a distribution associated with the relevant population. In order to estimate the relevance model we need a relevant example **r**. This appears to lead us into the standstill faced by previous researchers: in most retrieval scenarios we are given no examples of relevant documents. At this point we make use of the generative hypothesis: according to the GRH the user's query **q** may be viewed as a sample from the relevant population. Accepting the hypothesis gives us a justification for estimating a relevance model $RM_{\mathbf{q}}$ based on the user's query. The fact that **q** is not a document is irrelevant, since we view both documents and queries as projections from the same underlying representation space. Another cause for concern is the small size of **q**: typical queries consist of only a few words, which will certainly affect the quality of our estimate. However, since relevance models are Bayesian estimates we will be able to construct reasonable-looking and good-performing distributions even from 2-word samples. Once $RM_{\mathbf{q}}$ is estimated, we can use it to compute $P(\mathbf{d}|R{=}1)$, the probability of some document **d** under the model of the relevant class.

The second component of the probability ranking principle is the model of the non-relevant class $P(\mathbf{d}|R{=}0)$. Fortunately, this model is very easy to construct: from experience we know that the vast majority of documents in any given collection will not be relevant to the user's information need.[3] Accordingly, word statistics from the entire collection $\mathcal{C}$ will form a good approximation to the statistics of the non-relevant class. We will refer to these collection-wide word statistics as the *background* model BG:

$$BG(v) = \frac{n(v,\mathcal{C})}{\sum_v n(v,\mathcal{C})} \tag{3.15}$$

Here v is a word from the underlying vocabulary (we are presenting the simplest, mono-lingual case), and $n(v,\mathcal{C})$ is the total number of times v occurs in the collection. The denominator reflects aggregate collection size. Note that in a multi-lingual setting we would define a separate background model for each language involved. We would like to point out a connection between the background model and the relevance models discussed in section 3.5: $BG(\cdot)$ is a equivalent to a relevance model $RM_{\mathbf{r}}(\cdot)$ when the relevant sample **r** is the empty string. From a Bayesian perspective, the background model uses the prior distribution $p(\theta)$ in place of the relevance-based posterior $p(\theta|\mathbf{r})$ in equation (3.10).

Given the choices we made above, the probability ranking principle will assign the following score to a document $\mathbf{d} = d_1 \ldots d_n$:

[3] This excludes very broad "topics", such as the categories typically used in text classification research. Arguably, such topics are too coarse-grained to represent an information need.

$$P(R=1|\mathbf{d}) \propto \frac{P(\mathbf{d}|R=1)}{P(\mathbf{d}|R=0)} \approx \frac{\prod_{i=1}^{n} RM_{\mathbf{q}}(d_i)}{\prod_{i=1}^{n} BG(d_i)} \tag{3.16}$$

Here we are using the query-based relevance model $RM_{\mathbf{q}}$ to approximate the law of the relevant class and the background model BG as the law of the non-relevant class. Individual words d_i in the document $\mathbf{d}$ are assumed to be mutually independent given the value of the relevance variable R – this is consistent with the way PRP has been traditionally defined [117] and used for ranking.

3.6.2 Retrieval as hypothesis testing

In section 3.1.3, we outlined a retrieval approach centered around the idea of directly testing the hypothesis of relevance for each document in the collection. In this section we will provide specific definitions for this *hypothesis-testing* approach and relate it to several well-known ranking formulas. Recall that for a given query $\mathbf{q}$ and document $\mathbf{d}$ we define two hypotheses:

H_R : $\mathbf{q}$ and $\mathbf{d}$ were drawn from the same population
H_0 : $\mathbf{q}$ and $\mathbf{d}$ were drawn from unrelated populations

Under the GRH, H_R is equivalent to stating that $\mathbf{d}$ is relevant to $\mathbf{q}$, while H_0 means that $\mathbf{d}$ is not relevant. Accordingly, if we are able to reject the null hypothesis in favor of H_R, we have every reason to believe that $\mathbf{d}$ is in fact relevant. When presented with a collection of documents $\mathcal{C}$, we should rank the documents by the probability of rejecting the null hypothesis H_0. Furthermore, this form of ranking can be shown to be equivalent to Robertson's probability ranking principle[114], since GRH equates the event underlying H_R with the event R=1. Note that equivalence to PRP does not imply equivalence to the *probability ratio*, which was discussed in the previous section and is a special case of PRP.

The first issue we need to resolve is selecting the type of hypothesis test to be used for comparing H_R against H_0. There exist a wide variety of statistical methods for testing whether two random samples originated in the same population. A detailed comparison of these methods is beyond the scope of this book, but we will briefly discuss two applicable tests from the popular *likelihood ratio* family that are particularly well-suited to our domain: the sample likelihood test and Kullback's variation of Pearson's χ^2 test.

Sample likelihood test

The general idea of the sample likelihood test is to consider the ratio of probabilities for the two competing hypotheses, conditioned on the data that was observed. In our case the observation is a query-document pair $\mathbf{q}, \mathbf{d}$, and the competing hypotheses are H_R and H_0, so the ratio takes the following form:

$$\Lambda(\mathbf{q},\mathbf{d}) = \frac{P(H_R|\mathbf{q},\mathbf{d})}{P(H_0|\mathbf{q},\mathbf{d})} = \frac{P(\mathbf{q},\mathbf{d}|H_R)}{P(\mathbf{q},\mathbf{d}|H_0)} \cdot \frac{P(H_R)}{P(H_0)} \tag{3.17}$$

The second step in equation (3.17) comes from applying Bayes' theorem, with $P(H_0)$ and $P(H_R)$ representing the a-priori likelihoods of the competing hypotheses. Now let us consider the ratio on the right-hand side. The numerator $P(\mathbf{q},\mathbf{d}|H_R)$ represents the probability of observing $\mathbf{q},\mathbf{d}$ under the *relevant* hypothesis, i.e. as a sample from the same population. The denominator $P(\mathbf{q},\mathbf{d}|H_0)$ is the probability of $\mathbf{q}$ and $\mathbf{d}$ being drawn from two unrelated populations. This observation allows us to re-write the likelihood ratio in a more specific fashion, incorporating our model for observing the strings $\mathbf{q} = q_1 \ldots q_m$ and $\mathbf{d} = d_1 \ldots d_n$:

$$\begin{aligned}\frac{P(\mathbf{q},\mathbf{d}|H_R)}{P(\mathbf{q},\mathbf{d}|H_0)} &= \frac{P(\mathbf{q},\mathbf{d})}{P(\mathbf{q})P(\mathbf{d})} \\ &= \frac{\int_\Theta \left[\prod_1^m P_\theta(q_i)\right]\left[\prod_1^n P_\theta(d_j)\right] p(\theta)d\theta}{\left[\int_\Theta \prod_1^m P_\theta(q_i)p(\theta)d\theta\right]\left[\int_\Theta \prod_1^n P_\theta(d_j)p(\theta)d\theta\right]}\end{aligned} \tag{3.18}$$

The first step above follows directly from the definitions of our hypotheses: H_R implies that the query and the document were drawn together, as a single sample, while H_0 assumes that they were drawn independently. The second step in equation (3.18) comes from using equation (3.1) to express the probabilities of observing $\mathbf{q},\mathbf{d}$ as a single sample, or $\mathbf{q}$ and $\mathbf{d}$ as two unrelated samples. We can get an intuitive explanation for the cumbersome-looking integrals if we regard the parameter vector θ as an unknown underlying population. In the numerator, all the words $q_1 \ldots q_m, d_1 \ldots d_n$ are drawn independently, but from the same population θ. The integral reflects our uncertainty about that common population. In the denominator, we draw $q_1 \ldots q_m$ from one uncertain population, and then draw $d_1 \ldots d_n$ from another, completely unrelated population.

The test statistic $\Lambda(\mathbf{q},\mathbf{d})$ can be used directly for ranking the documents $\mathbf{d}$ in the collection. Higher values of the test statistic indicate a higher likelihood that $\mathbf{d}$ is in fact relevant to the user's need. Theoretically, it is also possible to convert $\Lambda(\mathbf{q},\mathbf{d})$ into an actual significance value (p-value), which could be used to conclusively decide whether H_0 should be rejected in favor of H_R. However, the conversion is not entirely straightforward, and will be left for future exploration.

Connection to the probability ranking principle

Before we turn our attention to the χ^2 test, we would like to highlight an interesting connection between the sample likelihood test and the probability ranking principle discussed in section 3.6.1. A careful examination of equation (3.18) reveals the three components which together define the posterior density of a relevance model, which was originally defined in equation (3.9), and is repeated here for reading convenience:

$$p(\theta|\mathbf{q}) = \frac{\left[\prod_{i=1}^{m} P_\theta(q_i)\right] p(\theta)}{\int_\Theta \left[\prod_{i=1}^{m} P_\theta(q_i)\right] p(\theta) \mathrm{d}\theta} \tag{3.19}$$

This observation allows us to re-write the sample likelihood statistic in the following form:

$$\Lambda(\mathbf{q}, \mathbf{d}) = \frac{\int_\Theta \left[\prod_{j=1}^{n} P_\theta(d_j)\right] p(\theta|\mathbf{q}) \mathrm{d}\theta}{\int_\Theta \left[\prod_{j=1}^{n} P_\theta(d_j)\right] p(\theta) \mathrm{d}\theta} \cdot \frac{P(H_R)}{P(H_0)} \tag{3.20}$$

This final form of $\Lambda(\mathbf{q}, \mathbf{d})$ allows us to make an interesting comparison with the probability ranking principle discussed in section 3.6.1. If we move the integrals inside the products, we will get precisely the probability ratio that forms the basis of the probability ranking principle:

$$\begin{aligned} \frac{\int_\Theta \left[\prod_{j=1}^{n} P_\theta(d_j)\right] p(\theta|\mathbf{q}) \mathrm{d}\theta}{\int_\Theta \left[\prod_{j=1}^{n} P_\theta(d_j)\right] p(\theta) \mathrm{d}\theta} &\rightarrow \frac{\prod_{j=1}^{n} \int_\Theta P_\theta(d_j) p(\theta|\mathbf{q}) \mathrm{d}\theta}{\prod_{j=1}^{n} \int_\Theta P_\theta(d_j) p(\theta) \mathrm{d}\theta} \\ &\approx \frac{\prod_{j=1}^{n} RM_{\mathbf{q}}(d_j)}{\prod_{j=1}^{n} BG(d_j)} \\ &\approx \frac{P(\mathbf{d}|R{=}1)}{P(\mathbf{d}|R{=}0)} \end{aligned} \tag{3.21}$$

In other words, the probability ratio advocated by Robertson [114] can be viewed as a sample likelihood test, provided that a relevance model is used in place of $P(\cdot|R{=}1)$. The only difference between the test and the ratio is that the latter assumes word independence and uses a point estimate, whereas the former is based on exchangeability and a complete Bayesian estimate. We must also point out that $\Lambda(\mathbf{q}, \mathbf{d})$ is substantially more expensive from a computational standpoint, unless the density $p(\theta)$ is selected so as to avoid integration at retrieval time.

χ^2 test of the minimum discrimination information statistic

The second hypothesis test we consider will be based on a very different idea. Instead of directly measuring the likelihood of H_0 against H_R (as in the ratio test), we will derive a p-value from a test statistic that is known to be χ^2-distributed. Specifically, we will test whether two observations $\mathbf{q}$ and $\mathbf{d}$ were drawn from the same population by comparing the posterior distributions they induce over the underlying space of events.

Let $\mathcal{V}$ be a vocabulary and let $RM_{\mathbf{q}}(\cdot)$ and $RM_{\mathbf{d}}(\cdot)$ denote the relevance models estimated from the query and the document respectively. Both of these relevance models are probability distributions over the events (words) in $\mathcal{V}$. When operating in multi-modal environments, we can always choose a single language and estimate relevance models over it as described in section 3.5.3. Suppose we draw N words at random from the query-based distribution $RM_{\mathbf{q}}(\cdot)$ and observe the counts $N(v)$ for every word v in the vocabulary.[4] Kullback and co-authors [66] define the *minimum discrimination information statistic (m.d.i.s.)* to be the following quantity:

$$\lambda(\mathbf{q},\mathbf{d}) = 2\sum_{v\in\mathcal{V}} N(v)\ln\left(\frac{N(v)}{N\cdot RM_{\mathbf{d}}(v)}\right) \tag{3.22}$$

The key property of m.d.i.s. is that asymptotically, as N grows, the distribution of $\lambda(\mathbf{q},\mathbf{d})$ values will approach a χ^2 distribution with α degrees of freedom, where α is one less than the number of words with non-zero counts $N(v)$. Under the *null* hypothesis H_0, which assumes that the query and the document were drawn from unrelated populations, $\lambda(\mathbf{q},\mathbf{d})$ follows a *non-central* χ^2 distribution; under H_R the distribution is mean-adjusted (centered). Neyman and Pearson [97] provide a detailed proof of the above claim, and Wilks [148] suggests a slightly different high-level argument for the same fact. The immediate consequence of the proof is that we can compute a p_0 value, which represents our confidence in the *null* hypothesis H_0 as:

$$p_0 = P(\chi^2_\alpha \le \lambda(\mathbf{q},\mathbf{d})) = \frac{\gamma(\frac{\alpha}{2}, \frac{\lambda(\mathbf{q},\mathbf{d})}{2})}{\Gamma(\frac{\alpha}{2})} \tag{3.23}$$

Here $\Gamma(a)$ represents the gamma function and $\gamma(a,x)$ is the lower incomplete gamma function.[5] The p_0 value given by equation (3.23) can be used in the same manner as with any statistical hypothesis test: if p_0 is sufficiently small, we are compelled to refute the null hypothesis and conclude that $\mathbf{d}$ and $\mathbf{q}$ were drawn from the same population, hence the document is relevant. Equation (3.23) can also be used for ranked retrieval: arranging the documents $\mathbf{d}$ in order of *increasing* p_0, should prove to be optimal in the same sense as Robertson's PRP [114].

Connection to language-modeling approaches

A reader familiar with information theory will no doubt have noticed a curious similarity between m.d.i.s. and Kullback-Leibler divergence, a popular measure of distance between two probability distributions. Indeed, if our sample size N is sufficiently large, relative word frequencies $\frac{N_v}{N}$ will closely approximate $RM_{\mathbf{q}}(v)$, which allows us to re-express $\lambda(\mathbf{q},\mathbf{d})$ as a function of KL divergence between $RM_{\mathbf{q}}$ and $RM_{\mathbf{d}}$:

$$\lambda(\mathbf{q},\mathbf{d}) = 2\sum_{v\in\mathcal{V}} N\cdot RM_{\mathbf{q}}(v)\ln\left(\frac{RM_{\mathbf{q}}(v)}{RM_{\mathbf{d}}(v)}\right)$$

---

[4] $N(v)$ are allowed to take zero values; these do not adversely affect the m.d.i.s.

[5] The lower incomplete gamma function is defined as $\gamma(a,x) = \int_0^x t^{a-1}e^{-t}dt$.

$$= \frac{2N}{\log_2 e} KL(RM_{\mathbf{q}}||RM_{\mathbf{d}}) \tag{3.24}$$

The $\frac{1}{\log_2 e}$ term appears because KL divergence is an information-theoretic quantity measured in *bits*, hence it is usually defined using a base-2 logarithm, whereas m.d.i.s. uses natural logs. Equation (3.24) allows us to draw interesting parallels between m.d.i.s. hypothesis testing and several existing retrieval models. Before we highlight these connections, we stress two important facts about the m.d.i.s. statistic:

(i) our formulation of m.d.i.s. treats the query $\mathbf{q}$ as a reference observation and the document $\mathbf{d}$ as a model that is either consistent or inconsistent with that observation. Since GRH is symmetric in the sense that documents and queries are viewed as objects of the same kind, it is equally possible to treat the document as a reference observation, using $KL(RM_{\mathbf{d}}||RM_{\mathbf{q}})$ instead of $KL(RM_{\mathbf{q}}||RM_{\mathbf{d}})$. The terms *document generation* and *query generation* are sometimes used to denote these two orientations.

(ii) we used relevance-model estimates $RM_{\mathbf{q}}$ and $RM_{\mathbf{d}}$ to represent word distributions associated with the query and the document. While relevance models arise naturally from the generative hypothesis, the m.d.i.s. statistic can be defined using *any* two distributions that reflect the underlying populations of $\mathbf{q}$ and $\mathbf{d}$. For example, a maximum-likelihood distribution like $\frac{n(w,\mathbf{d})}{|\mathbf{d}|}$ can be used as a reference sample on the left side of the KL divergence. The only restriction is that we cannot allow zero-probability events on the right side of KL.

Keeping in mind the above observations, we are ready to interpret several existing retrieval models in terms of a hypothesis-testing approach.

1. **Simple language models.** Recall from section 2.3.4 that a language-modeling approach involves ranking the documents by the probability of observing the query $\mathbf{q}$ as a sample from the document model $M_{\mathbf{d}}(v)$, estimated as the smoothed relative frequency of each word v in the document: $\frac{n(v,\mathbf{d})+\mu BG(v)}{|\mathbf{d}|+\mu}$. It is easy to show that the scoring formula used by the query likelihood model is rank-equivalent to the m.d.i.s. statistic:

$$\begin{aligned} P(\mathbf{q}|M_{\mathbf{d}}) &= \prod_{i=1}^{|\mathbf{q}|} M_{\mathbf{d}}(q_i) \\ &= \exp\left(\sum_{v\in\mathcal{V}} n(v,\mathbf{q}) \ln M_{\mathbf{d}}(v)\right) \\ &\stackrel{\text{rank}}{=} 2\sum_{v\in\mathcal{V}} n(v,\mathbf{q}) \ln \frac{M_{\mathbf{d}}(v)|\mathbf{q}|}{n(v,\mathbf{q})} \\ &= -\lambda(\mathbf{q},\mathbf{d}) \end{aligned} \tag{3.25}$$

The consequence of equation (3.25) is that query-likelihood ranking can be interpreted as running a series of statistical tests to determine which documents in the collection were sampled from the same population as the query $\mathbf{q}$. Furthermore, combining equations (3.25) and (3.23) will give us a p_0 value, which can be used to accept or reject the hypothesized relevance of any given document. The main difference from our approach is that simple frequency-based estimates are used instead of $RM_{\mathbf{q}}$ and $RM_{\mathbf{d}}$.

2. **Mixture-based models.** The next family of models we consider includes cluster-based language models [152, 67], aspect-based models [58] and retrieval models inspired by ideas from machine translation [11, 153]. What differentiates these models from the simple query-likelihood approach is their use of a sophisticated mixture-based estimate to represent the document-based distribution $M_{\mathbf{d}}$:

$$M_{\mathbf{d}}(v) = \sum_z P(v|z)P(z|\mathbf{d}) \tag{3.26}$$

The meaning of variable z above varies from model to model. In the *cluster-based* models, z represents a group of documents, obtained either by partitioning the collection [152], or by constructing nearest-neighbor cliques around every document $\mathbf{d}$ [67]. The quantity $P(v|z)$ represents a language model based around the cluster z and is typically constructed by averaging simple frequency-based models of every document in the cluster. In *aspect-based* models [58], z represents a latent dimension and the language model $P(v|z)$ is estimated using the EM algorithm as will be discussed in section 4.2.4. In *translation-based* models [11, 153], z represents a word in the vocabulary and $P(v|z)$ represents the probability of translating z into v. What all of these models have in common is that after $M_{\mathbf{d}}$ is estimated, the documents are ranked using the query-likelihood criterion $P(\mathbf{q}|M_{\mathbf{d}})$. As we discussed above, query-likelihood is rank-equivalent to m.d.i.s., except that a simple frequency-based estimate is used in place of $RM_{\mathbf{q}}$. A p_0-value derived from $P(\mathbf{q}|M_{\mathbf{d}})$ can be used to accept or reject the null hypothesis for a given document.

3. **KL-divergence models.** The final family of models we consider represents a confluence of query-likelihood ideas discussed above and document-likelihood ideas inherent in the probability ranking principle. Query-likelihood research is typically focused on producing accurate estimates of *document models* $M_{\mathbf{d}}$, while the query $\mathbf{q}$ is treated as an observation. Document-likelihood approaches are concerned with estimating a *query model* $M_{\mathbf{q}}$, which is then used to compute the probability of observing documents $\mathbf{d}$. Croft [33] was the first to suggest that a superior form of ranking may involve estimating both the query model and the document model, with the purpose of directly comparing the two. Zhai and Lafferty [68, 157] used the same idea in the development of the highly-successful risk-minimization framework for information retrieval. Retrieval

approaches in this family involve estimating $M_{\mathbf{q}}$ and $M_{\mathbf{d}}$, and then measuring distance between the two models using the Kullback-Leibler divergence. The query model $M_{\mathbf{q}}$ is estimated in a non-trivial fashion, often using pseudo-feedback techniques very similar to *relevance models*. The document model $M_{\mathbf{d}}$ is typically based on smoothed relative frequencies, but can be estimated in a cluster-based fashion [67]. The divergence $KL(M_{\mathbf{q}}||M_{\mathbf{d}})$ is usually interpreted in an information-theoretic sense, as a number of bits wasted when encoding the query model with the document model, however Zhai [157] provides a much deeper interpretation as a decision-theoretic notion of *risk* associated with presenting **d** to the user. It should be obvious that any model based on KL divergence has an immediate interpretation in terms of the m.d.i.s. hypothesis test: the ranking formulae are virtually identical. However, computing the significance value p_0 is not as straightforward as with the query-likelihood models. The reason is that KL-based models employ a query model $M_{\mathbf{q}}$ in place of the actual sample **q**, which gives us probabilities $M_{\mathbf{q}}(v)$ instead of observed word counts $n(v, \mathbf{q})$. While equation (3.24) can be used to compute the m.d.i.s. statistic, the sample size N will need to be set in a heuristic fashion and may artificially inflate significance values.[6] We must note that the same deficiency is present in our formulation, which uses relevance models $RM_{\mathbf{q}}$ and $RM_{\mathbf{d}}$ to represent query and document populations. Resolving this deficiency is left for future work.

3.6.3 Probability ratio or KL-divergence?

In the previous two sections we presented two general ways of ranking documents with relevance models. The first (section 3.6.1 was based on the probability ranking principle [114] and arranged the documents in the decreasing order of the probability ratio: $\frac{P(\mathbf{d}|R=1)}{P(\mathbf{d}|R=0)}$. The second (section 3.6.2) was motivated by the generative relevance hypothesis and ranked the documents by the probability that they were sampled from the same population as the query, which was shown to be rank-equivalent to $KL(RM_{\mathbf{q}}||RM_{\mathbf{d}})$. In this section we will attempt an analytic comparison of these two ranking methods with the aim of determining which method may lead to better retrieval performance.

Let us assume for the moment that our relevance model $RM_{\mathbf{q}}$ is the best possible estimate for the model of the relevant class. That means our estimate $RM_{\mathbf{q}}(v)$ accurately reflects $P(v|R=1)$, the actual probability of word

[6] As N increases to infinity in equation (3.24), the significance value p_0 will almost surely approach one, except for the unlikely case when the document and the query are identical. At the same time, N must be larger than α: a statistical paradox will arise if the aggregate number of observed events is fewer than the degrees of freedom for the χ^2 test. Since models $M_{\mathbf{q}}$ and $M_{\mathbf{d}}$ provide non-zero mass for every word in the vocabulary, we have $\alpha=|\mathcal{V}|-1$ which means that N has to be at least as large as our vocabulary size.

v appearing in documents relevant to the user's need. We will argue that in this case, using KL divergence may result in better ranking than using the probability ratio. In order to support this claim, we are going to consider the following question:

> *Out of all possible documents* $\mathbf{d}$*, which one would be ranked the highest by each method?*

First let us consider the probability ratio. We will initially assume that all documents under consideration have the same number of words n; the assumption will later be relaxed. In mathematical terms, we are looking for a vector of word frequencies $\mathbf{d}(v)$ which would maximize the probability ratio $\frac{P(\mathbf{d}|R=1)}{P(\mathbf{d}|R=0)}$ subject to the constraint $\sum_v \mathbf{d}(v) = n$. For any given document $\mathbf{d}$ of length n the following inequality holds:

$$\frac{P(\mathbf{d}|R=1)}{P(\mathbf{d}|R=0)} = \prod_{v \in \mathcal{V}} \left(\frac{P(v|R=1)}{P(v|R=0)}\right)^{\mathbf{d}(v)}$$

$$\leq \left(\frac{P(v^*|R=1)}{P(v^*|R=0)}\right)^n \times \prod_{v \neq v^*} \left(\frac{P(v|R=1)}{P(v|R=0)}\right)^0 \tag{3.27}$$

where $v^* = \arg\max_v \left\{\frac{P(v|R=1)}{P(v|R=0)}\right\}$. In other words, to get the highest possible probability ratio score, the document $\mathbf{d}$ should contain n repetitions of a single word v^*. The magic word v^* should have a high probability $P(v|R=1)$ of occurring in the relevant documents and a comparatively low probability $P(v|R=0)$ of occurring in the non-relevant set. Now we can relax the assumption made earlier that every document should contain exactly n words. Since $P(\cdot|R=1)$ and $P(\cdot|R=0)$ are probability distributions we have $\sum_v P(v|R=1) = \sum_v P(v|R=0)$, and therefore $\frac{P(v^*|R=1)}{P(v^*|R=0)} \geq 1$. It should be evident that repeating v^* more and more times will only increase the probability ratio of the optimal document $\mathbf{d}$, since multiplying by a number greater than 1 will never decrease the score. The consequence is that as far as probability ratio is concerned, the perfect document $\mathbf{d}^*$ would contain a single word v^*, repeated many times in a row. Of course, documents repeating a single word are not likely to occur in any realistic collection. However, the implication of our argument is this:

> *Probability ratio will favor documents that contain many repetitions of a few highly discriminative words, i.e. words with large values of* $\frac{P(v|R=1)}{P(v|R=0)}$.

The argument above is not limited to probability-ratio scoring. The reasoning holds for any entropy-based ranking formula where the document is treated as an observation, for example $KL(M_{\mathbf{d}}||M_{\mathbf{q}})$.

Now we will demonstrate that KL-divergence will favor a very different kind of document. As before, we want to find the document $\mathbf{d}$ which will achieve the highest possible score $-KL(RM_{\mathbf{q}}||M_{\mathbf{d}})$. As before, $RM_{\mathbf{q}}(v)$ is the best available approximation to $P(v|R{=}1)$, and $M_{\mathbf{d}}$ is the document model based on relative frequencies of words.[7] For any document $\mathbf{d}$ we have:

$$\begin{aligned} -KL(M_{\mathbf{q}}||M_{\mathbf{d}}) &= \sum_{v\in\mathcal{V}} RM_{\mathbf{q}}(v) \log \frac{M_{\mathbf{d}}(v)}{RM_{\mathbf{q}}(v)} \\ &\leq \log\left(\sum_{v\in\mathcal{V}} RM_{\mathbf{q}}(v) \frac{M_{\mathbf{d}}(v)}{RM_{\mathbf{q}}(v)}\right) \\ &= 0 \;=\; KL(RM_{\mathbf{q}}||RM_{\mathbf{q}}) \end{aligned} \tag{3.28}$$

The second step in equation (3.28) is Jensen's inequality applied to the expectation of a logarithm; the last step comes from the fact that relative entropy of a distribution with itself is zero. The meaning of equation (3.28) is that the highest-ranking document $\mathbf{d}^*$ will distribute its probability mass among words in the same way as the relevance model $RM_{\mathbf{q}}$. As before, such a document $\mathbf{d}^*$ may not exist in our collection, but the implication of our argument is this:

> *KL-divergence will favor documents that contain many distinct words which are prevalent in the relevant class, i.e. words for which $P(v|R{=}1)$ is non-zero.*

Again, the argument will apply to any entropy-based ranking formula which treats the query as reference observation, such as the query-likelihood $P(\mathbf{q}|M_{\mathbf{d}})$ or cross-entropy $H(M_{\mathbf{q}}||M_{\mathbf{d}})$.

The conclusion of our analysis is that probability ratio and KL-divergence represent two very different ranking principles. They will favor two very different types of documents: the former will prefer a few highly relevant words, the latter will opt for as many possibly relevant words as possible. Our intuition suggests that KL-divergence ought to be the better choice: it will tend to select documents $\mathbf{d}$ that look like the expected relevant document. However, empirical performance being the ultimate judge, we will test both alternatives in chapter 5.

3.6.4 Summary of ranking methods

In this section we discussed how relevance models could be used for ranking documents in the collection. We focused on two formalisms: the well-known probability ranking principle and a novel idea of treating information retrieval as hypothesis testing. The application of the probability ranking principle, discussed in section 3.6.1, was fairly straightforward: we used the query-induced

[7] $-KL(RM_{\mathbf{q}}||M_{\mathbf{d}})$ represents a typical state-of-the-art instantiation of the KL-divergence model. $RM_{\mathbf{d}}$ is less frequent due to the difficulty of indexing non-sparse language models

relevance model $RM_{\mathbf{q}}$ to represent the relevant population and a collection-wide distribution BG to model the non-relevant population. The hypothesis-testing approach to information retrieval was formalized in section 3.6.2. The basic idea of the approach is to directly test the hypothesis that the document $\mathbf{d}$ was drawn from the same underlying population as the query $\mathbf{q}$. We discussed two appropriate significance tests: the sample likelihood test, and the χ^2 test on the m.d.i.s. statistic.

We have also discussed the relationships between our hypothesis-testing ideas and several existing retrieval models. Specifically, we demonstrated that our application of Robertson's probability ranking principle is strongly related to the sample likelihood test. We have also shown that a number of popular language-modeling approaches can be interpreted as variations of the χ^2 test on the m.d.i.s. statistic, including the simple query-likelihood model, mixture-based language models and the KL-divergence model that serves as the basis for the risk minimization framework. Finally, in section 3.6.3 we provided an analytical comparison of two main ranking formulas: the probability ratio central to the probability ranking principle, and the Kullback-Leibler divergence, which forms the basis for the hypothesis-testing approach. We have demonstrated that the two formulas will favor very different types of documents and suggested that a KL-based formula may be superior.

The multitude of ranking formulas discussed in this section may appear confusing, but it is important to realize that they are strongly related to each other and differ only in the way they represent the query model $M_{\mathbf{q}}$ or the document model $M_{\mathbf{d}}$. In our experience, a hypothesis test based on $KL(RM_{\mathbf{q}}||RM_{\mathbf{d}})$ will generally provide ranking performance superior to other formulations we discussed. However, other considerations may influence the choice of estimates for the query and document models. For example, using a relevance model $RM_{\mathbf{d}}$ to represent documents substantially complicates the process of constructing an inverted index. Accordingly, many of our experiments use a frequency-based estimate $M_{\mathbf{d}}$ for documents and a relevance model $RM_{\mathbf{q}}$ for queries. In multi-media scenarios, the opposite is often true: indexing the content of an image or a video frame is often more challenging than indexing a text-valued relevance model $RM_{\mathbf{d}}$. Finally, a domain like Topic Detection and Tracking presents its own unique challenges, for which best performance is obtained using a formula derived from the probability ratio.

3.7 Discussion of the Model

This chapter was dedicated to the theoretical development of the generative view of relevance. We provided formal definitions of the four components of our model: **(i)** the representation of documents, queries and relevance, **(ii)** the probabilistic distributions over documents and queries, **(iii)** distributions associated with the relevant population (*relevance models*), and **(iv)** the criteria for ranking documents in response to a user's query. In our discussion we skipped one seemingly small component: we never specified the probability distribution $p(\theta)$ over the parameter space Θ. This omission is by no means accidental. In fact, $p(\theta)$ plays an absolutely critical part in our model, and we felt it necessary to devote the entire next chapter to its development. In the present chapter we were also guilty of keeping our definitions unreasonably abstract. We did not fully specify the nature of the information space $\mathcal{S}$, and neither did we talk about the meaning of random variables $X_1 \ldots X_M$ that form the latent representation of documents and queries. This too was done on purpose – we wanted to keep our definitions as general as possible, so the model would be applicable to a wide range of specific retrieval scenarios. We will provide the complete representation details in chapter 5. We will now take some time to discuss the advantages of our model and highlight differences from previously proposed retrieval models.

The centerpiece of our model is the generative relevance hypothesis. To the best of our knowledge, none of the existing retrieval models make a similar assumption about queries and relevant documents. The classical model of Robertson and Sparck Jones [117] does assume a common model for relevant documents, but it never considers that user queries could be treated as samples from the same model. On the other hand, the language-modeling approach of Ponte and Croft [106] tests whether the query could be a sample from a *given* document model, but it never conceives that *all* relevant documents might be samples from a common model. Other probabilistic formulations either follow the classical model [117] or the language-modeling paradigm [106]. One noteworthy exception is the *risk-minimization framework* of Zhai and Lafferty [68], where the authors hypothesize the existence of *two* generative models: one is responsible for generating documents, the other generates queries. There is no overlap between the two models: they do not have to share the same event space, and the process of generating documents can be completely different than the process of sampling queries. The fact that query generation and document generation are independent processes allows greater freedom in the choice of estimation methods, but also introduces a challenge similar to the one encountered during the development of the *unified probabilistic model* [115, 116]: documents and queries live in different, non-overlapping parameter spaces, which makes defining relevance non-trivial. Zhai and Lafferty circumvent the challenge by removing relevance altogether and replacing it with a decision-theoretic notion of *risk*, which serves as a bridge between the two unrelated generative processes.

We believe our model represents the first attempt to tie together in a completely formal fashion all three components of a retrieval system: documents, queries and relevance. Recall that the classical model and its extensions generally focus on the link between relevance and documents. The query exists only as a set of heuristics to help us estimate relevant probabilities, it is never treated as an observation. To contrast that, language modeling approaches focus exclusively on the link between documents and queries. Relevance is not a formal part of the model, and as we discussed in section 2.3.4, introducing it may not be a trivial matter. Tying both documents and queries to relevance allows us to avoid heuristics even when no relevant examples are provided. When we do have relevant examples, we treat them just like the query – as samples from the underlying relevance model.

Almost every probabilistic model of information retrieval assumes word *independence* in one form or another. The difference in our model is in that we assume *exchangeability* rather than independence. Does this matter? In light of our discussion in sections 2.3.2 and 2.3.3 one may be tempted to say that there is no benefit in modeling dependence. We beg to differ. Exchangeability, coupled with our estimates for the generative density $p(\theta)$ will allow us to do things that cannot be easily done in other models. It will allow us to estimate relevant probabilities using the query alone. It will make massive query expansion a formal part of the model. It will allow us to associate English queries with Chinese documents, let us annotate a video segment with appropriate keywords, or retrieve a bitmap image based on a text query. In our opinion, assuming exchangeability instead of independence may prove to be quite beneficial.

4

Generative Density Allocation

The present chapter plays a special role in this book. In this chapter we will not be talking about relevance, documents or queries. In fact, this chapter will have very little to do with information retrieval. The subject of our discussion will be generative models for collections of discrete data. Our goal is to come up with an effective generative framework for capturing *interdependencies* in sequences of exchangeable random variables. One might wonder why a chapter like this would appear in a book discussing relevance. The reason is simple: a generative model lies at the very heart of the main assumption in our model. Our main hypothesis is that there exists a generative model that is responsible for producing both documents and queries. When we construct a search engine based on the GRH, its performance will be affected by two factors. The first factor is whether the hypothesis itself is true. The second factor is how accurately we can estimate this unknown generative process from very limited amounts of training data (e.g. a query, or a single document). Assuming the GRH is true, the quality of our generative process will be the single most important influence on retrieval performance. When we assume the generative hypothesis, we are in effect reducing the problem of information retrieval to a problem of generative modeling. If we want good retrieval performance, we will have to develop effective generative models.

4.1 Problem Statement

The previous chapter laid the groundwork for the kinds of generative models we will be dealing with. We are interested in probability distributions for sequences of exchangeable random variables $X_1 \dots X_n$, each taking values in some event space $\mathcal{S}_i$. In order to make our discussion clearer we will make a following simplification. We assume that random variables X_i represent *words* from a common finite vocabulary $\mathcal{V}$. The sequences $X_1 \dots X_n$ represent strings or sentences over that vocabulary. Our goal is then to construct an effective model for finite strings of text, keeping in mind that order of the words in these strings does not matter. We would like to stress that we adopt the terms "words" and "text" only as a matter of convenience. The generative models we describe can be applied to any dataset of discrete exchangeable data. Furthermore, in chapter 5 we will show that our models can be generalized to real-valued variables. But for now, let us stick to strings of text, or, more accurately, *bags of words*, since the order of words in a string makes no difference.

4.1.1 Objective

As we mentioned above, our goal is to construct *effective* models of strings. By definition, a generative model is a probabilistic representation of the data. However, there may exist different points of view about what constitutes an effective probabilistic representation; we would like to make explicit what we are aiming for. We are not interested in *compressing* the data in either lossy or lossless fashion, so we do not care whether our representation is compact, or how many parameters it uses. Nor is it in our goals to come up with a grand theory for bags of words, so we are not going look for the most elegant representation. Our goal is to construct a *predictive* model that is able to generalize – i.e. a model that can be trained from a set of sample strings and then accurately predict new, previously unobserved strings. Specifically, we will assume that we have a training collection of strings $\mathcal{C}_{train}$, and a disjoint collection of testing strings $\mathcal{C}_{test}$. Our goal is to use strings from $\mathcal{C}_{train}$ to estimate all necessary parameters of the model, and then have the model predict every string from $\mathcal{C}_{test}$. A model that assigns comparatively higher likelihood to the testing strings will be considered superior, regardless of its form or the number of parameters it uses. We would like to stress that higher likelihood on $\mathcal{C}_{test}$ measures the ability of the model to *generalize* to new data, and not its ability to *overfit*.

4.2 Existing Generative Models

We will start our development by looking at five existing probabilistic models that are commonly applied to discrete exchangeable collections. We will consider consider the following models: **(i)** the unigram model, **(ii)** the mixture model, **(iii)** the Dirichlet model, **(iv)** probabilistic Latent Semantic Indexing (pLSI) and **(v)** Latent Dirichlet Allocation (LDA). The first two models are extremely simple and are frequently employed in a large number of fields. The Dirichlet model can be thought of as a Bayesian version of the unigram model. The last two models represent a somewhat less-known family of *simplicial mixture* models. Both are considerably more complex than the first three models, both require advanced methods of inference for parameter estimation. We will present the five models in order of increasing complexity. In each case we will identify a weak spot of the model – its inability to deal with some peculiarity of the data. Each subsequent model will address the weakness of its predecessor.

A reader familiar with language modeling will notice that our discussion does not include the following three families of models: (i) Markov chains, (ii) probabilistic grammars and (iii) maximum-entropy models. All three are considered to be staples of natural language processing. We do not discuss them because all three are based on the order of words in a string and cannot be applied to exchangeable sequences.

4.2.1 The Unigram model

The unigram model is perhaps the simplest model we can construct. The model consists of a single distribution U over our vocabulary $\mathcal{V}$. It is a vector of probabilities $U(v)$ one for each word v in the vocabulary. Informally we can think of U as an urn that contains $k{\cdot}U(v)$ counts of each word v, where k is some very large constant. To generate strings of text, we randomly pull out a word from this urn, observe its value, and put it back into the urn. The draws are independent of each other, so the probability of getting a particular sequence of words $w_1 \ldots w_n$ is:

$$P_{uni}(w_1 \ldots w_n) = \prod_{i=1}^{n} U(w_i) \tag{4.1}$$

Estimation of parameters for the unigram model is a very simple task. Given a collection of training strings $x \in \mathcal{C}_{train}$, we simply count how many times we observed each word v. The counts are then normalized and some form of smoothing is applied:

$$U(v) = \lambda \frac{\sum_x n(v,x)}{\sum_x |x|} + (1-\lambda) b_v \tag{4.2}$$

Here $n(v,x)$ denotes the number of times v occurred in the training string x, and $|x|$ stands for the length of x. Summations go over training string $x \in \mathcal{C}_{train}$. Smoothing is achieved by interpolating normalized frequencies with some *background* distribution b_v over the vocabulary. In the absence of prior information, b_v is usually assumed to be uniform over $\mathcal{V}$. The constant λ controls the degree of variance in the estimator. Setting $\lambda = 1$ turns equation (4.2) into the *maximum likelihood* estimator, i.e. the distribution U_{ml} which assigns the maximum possible probability to the training collection $\mathcal{C}_{train}$ under equation (4.1). U_{ml} also represents an *unbiased* estimator. By the law of large numbers, if we repeatedly estimate U_{ml} with random training collections $\mathcal{C}_{train}$, the quantity $U_{ml}(w)$ will eventually converge to the true probability for each word v. However, U_{ml} is a *high-variance* estimator, meaning the estimates $U_{ml}(v)$ will vary a lot if we consider different training collections, especially if these collections are relatively small. Setting $\lambda < 1$

will reduce the variance, but will also introduce a bias in the estimates, i.e. $U(v)$ will no longer converge to true probabilities.

The unigram model is simple, easy to estimate and is the first choice adopted by researchers dealing with discrete collections. However, it does have a serious drawback: it can only model *homogeneous* collections, where all strings are expected to be similar to each other. If we have two strings a and b, the unigram model would not be affected by taking some words from a and swapping them with words from b: the joint probability of a and b would be the same. This becomes a serious problem if string a discussed politics and string b talked about carburetors. We would like our generative model to recognize that political lingo is unlikely to be used in the same sentence as automotive jargon. This is particularly important because our ultimate interest lies in *topical* relevance of strings.

4.2.2 The Mixture model

The mixture model, also known as *cluster* model or *topic* model, represents a natural evolution of the unigram model. It is designed specifically for dealing with *heterogeneous* collections, where strings can talk about multiple topics. In the mixture model we assume that all possible strings fall into a finite set of k clusters. Strings in each cluster discuss a particular topic z, distinct from all other topics. Each topic z is associated with a distribution T_z over our vocabulary. Words v with high probabilities $T_z(v)$ reflect the jargon associated with topic z. The model also includes a mixture distribution $\pi(z)$, which tells us how prominent topic z is in the collection. When we want to generate a new string $w_1 \ldots w_n$ from the mixture model, we carry out the following process:

1. Pick a topic $z \in \{1 \ldots k\}$ with probability $\pi(z)$.
2. For $i = 1 \ldots n$: pick word w_i from topic z with probability $T_z(w_i)$

The overall likelihood of observing $w_1 \ldots w_n$ from the mixture model defined by parameters $\pi(\cdot)$ and $T_z(\cdot)$ is:

$$P_{mix}(w_1 \ldots w_n) = \sum_{z=1}^{k} \pi(z) \prod_{i=1}^{n} T_z(w_i) \tag{4.3}$$

Estimation of parameters in the mixture model is considerably more varied than in the unigram model. We are aware of two general approaches to estimation. The first is based on *clustering* the training strings, the second, more formal approach, is based on maximizing likelihood of $\mathcal{C}_{train}$. In the first approach (e.g. [152]) we cluster strings in the training collection $\mathcal{C}_{train}$ into k groups: $C_1 \ldots C_k$. The groups typically do not overlap, and result from applying any one of the wide range of clustering algorithms [82]. Once we group the strings, we can estimate a topic model T_z for each cluster C_z using equation (4.2), just as we did for the unigram model. However, it is also a common practice to interpolate relative word frequencies from cluster C_z with the overall unigram model U in the following fashion:

$$T_z(v) = \lambda_1 \frac{\sum_{x \in C_z} n(v,x)}{\sum_{x \in C_z} |x|} + \lambda_2 \frac{\sum_x n(v,x)}{\sum_x |x|} + \lambda_3 b_v \tag{4.4}$$

The first summation concerns the strings x that fall into the cluster of the current topic, while the second involves all training strings. The constants λ_1, λ_2 and λ_3, control the overall estimator variance; they should be positive and must add up to 1. The mixing distribution $\pi(z)$ is commonly taken to reflect the relative size of each cluster: $\frac{|C_z|}{|C_{train}|}$.

The second approach to parameter estimation is somewhat more formal. We assume that the training strings C_{train} represent a random sample from the generative model defined by equation (4.3), and try to find the distributions $\pi(\cdot)$ and $T_z(\cdot)$ which maximize the log-likelihood:

$$\sum_{x \in C_{train}} \log \left\{ \sum_{z=1}^{k} \pi(z) \prod_{i=1}^{n} T_z(w_i) \right\} \tag{4.5}$$

The parameters maximizing equation (4.5) cannot be expressed in closed form, but can be approximated using either iterative gradient ascent, or the Expectation Maximization (EM) algorithm. For details of the latter approach see [102].

The mixture model represents a definite improvement over the unigram model, but it has two important limitations. The first has to do with the number of topics. The number k has to be selected *a-priori*, before we start either the clustering or the maximization process. If we happen to pick a k that is too small, we will lose some resolution in the model: several distinct topics will be collapsed to the same distribution T_z, leading to the same problem that we encountered with the unigram model. Making k too large will lead to a different problem: our mixture model will have too many parameters, and the estimates will overfit to the training data.

The second limitation of the mixture model is somewhat more subtle. The model assumes that each string is generated by a single topic model T_z. Consequently, the mixture will assign unreasonably low probabilities to strings that discuss two or more distinct topics, e.g. sports and weather, or politics and medicine. This may not matter for short strings of text, such as single sentences, but leads to serious problems when we start dealing with complete news stories or editorials. In order to model these kinds of documents accurately, we would have to make k so large that it covers any combination of topics that might be prominent in the collection. And, as we mentioned above, making k too large will lead to serious over-fitting to the training data.

4.2.3 The Dirichlet model

The Dirichlet model is our first real example of Bayesian ideas in a generative model. We are going to devote slightly more space to it, because it will play an important role in the development of two subsequent models: Latent Dirichlet Allocation (LDA), and our own model. One way to view the Dirichlet model is as expression of *uncertainty* over the parameters U of the unigram model. Recall that U is a distribution over the vocabulary. At the same time, U is only a single point in the simplex $I\!P_{\mathcal{V}} = \{u{\in}[0,1]^{\mathcal{V}} : |u|{=}1\}$. The *simplex* is simply the set of all possible distributions over our vocabulary. When we adopted the unigram model we placed all our confidence on one point, U. From a Bayesian perspective that reflects extreme confidence in our estimate U. If we want to allow a bit of healthy uncertainty, we should spread our confidence over a number of points. That can be accomplished by introducing a density $p(u)$ over the points $u{\in}I\!P_{\mathcal{V}}$. Using $p(u)$ as a measure of uncertainty would give us the following process for generating a string of text $w_1 \ldots w_n$:

1. Pick a unigram estimate u with probability $p(u)$ from the distribution space $I\!P_{\mathcal{V}}$
2. For $i = 1 \ldots n$: pick word w_i from the unigram model u with probability $u(w_i)$

We should immediately recognize that the generative process for the Dirichlet model looks exactly like the process of the mixture model. The only difference is that before we were dealing with a finite number of topics $T_1 \ldots T_n$, whereas now we have an uncountably infinite space of topics $I\!P_{\mathcal{V}}$. In a way, the Dirichlet model can be viewed as a curious solution for deciding how many topics we should have: instead of deciding on k, just pick uncountably many. As a result, the summation of equation (4.3) will turn into an integral:

$$P_{dir}(w_1 \ldots w_n) = \int_{I\!P_{\mathcal{V}}} \left(\prod_{i=1}^{n} u(w_i) \right) p(u) du \tag{4.6}$$

A natural question might arise: if we had problems with parameter estimation for a finite number k of topics, how can we handle infinitely many of them? The reason is simple: in the mixture model we had to estimate an entire distribution T_z for every topic z. Here every topic u is already defined, we only have to pick the density $p(u)$. And in practice, considerations of computational complexity drive most researchers to pick a simple *conjugate prior* as the density $p(u)$. The appropriate conjugate prior for the unigram model is the *Dirichlet* distribution:

$$p(u) = \frac{\Gamma(\sum_{v \in \mathcal{V}} \alpha_v)}{\prod_{v \in \mathcal{V}} \Gamma(\alpha_v)} \prod_{v \in \mathcal{V}} u(v)^{\alpha_v - 1} \tag{4.7}$$

Here $\Gamma(x)$ is the Gamma function. The distribution has a set of $|\mathcal{V}|$ parameters α_v, one for each word v in the vocabulary. The parameters α_v can be estimated in closed form by treating the training strings $\mathcal{C}_{train}$ as an observation from equation (4.6) under the prior Dirichlet density where all parameters are set to a constant β. Omitting the derivation, the posterior estimates take a very simple form:

$$\alpha_v = \beta + \sum_{x \in \mathcal{C}_{train}} n(v,x) \tag{4.8}$$

Here $n(v,x)$ is the number of times the word v occurred in the training string x. Curiously, the total number of parameters in a Dirichlet model is the same as in the simple unigram model; one more to be exact, since α_v don't have to add up to one. The small number of parameters hints at the fact that the Dirichlet distribution may not be as complex as it seems. Indeed, we will show that the Dirichlet model *does not* contain an infinite number of topics. In fact it involves only one topic, slightly more sophisticated than the unigram model. If we plug the density $p(u)$ from equation (4.7) into equation (4.6), we will get:

$$\begin{aligned}
P_{dir}(w_1 \ldots w_n) &= \int_{\mathbb{P}_{\mathcal{V}}} \left(\prod_{i=1}^{n} u(w_i) \right) \left(\frac{\Gamma(\Sigma_v \alpha_v)}{\prod_v \Gamma(\alpha_v)} \prod_v u(v)^{\alpha_v - 1} \right) du \\
&= \frac{\Gamma(\Sigma_v \alpha_v)}{\prod_v \Gamma(\alpha_v)} \cdot \int_{\mathbb{P}_{\mathcal{V}}} \left(\prod_v u(v)^{\alpha_v - 1 + n(v, w_{1\ldots n})} \right) du \\
&= \frac{\Gamma(\Sigma_v \alpha_v)}{\prod_v \Gamma(\alpha_v)} \cdot \frac{\prod_v \Gamma(\alpha_v + n(v, w_{1\ldots n}))}{\Gamma(\Sigma_v \alpha_v + n)} \\
&= \frac{\Gamma(\Sigma_v \alpha_v)}{\Gamma(\Sigma_v \alpha_v) \cdot (\Sigma_v \alpha_v) \cdot (\Sigma_v \alpha_v + 1) \cdots (\Sigma_v \alpha_v + n - 1)} \\
&\quad \times \prod_v \frac{\Gamma(\alpha_v) \cdot (\alpha_v) \cdot (\alpha_v + 1) \cdots (\alpha_v + n(v, w_{1\ldots n}) - 1)}{\Gamma(\alpha_v)} \\
&= \prod_{i=1}^{n} \frac{\alpha_{w_i} + n(w_i, w_{1\ldots(i-1)})}{\Sigma_v \alpha_v + (i-1)}
\end{aligned} \tag{4.9}$$

If we squint really hard at equation (4.9), we will immediately recognize it as something very familiar. In fact, were it not for $n(w_i, w_{1\ldots(i-1)})$ on the top and $(i-1)$ on the bottom, equation (4.9) would be exactly the same as the unigram model (equation 4.1). The proportion $\frac{\alpha_{w_i}}{\Sigma_v \alpha_v}$ would play the role of $U(w_i)$, the probability of observing the word w_i from the unigram urn. What makes the Dirichlet model different is the component $n(v, w_{1\ldots(i-1)})$, which represents the number of times the current word w_i appears among previous words $w_1 \ldots w_{i-1}$. The Dirichlet model captures a very interesting kind of dependency: the dependency of a word on its own previous occurrences. Every appearance of a word v in the string makes subsequent appearances more likely. In this way the Dirichlet model reflects a very important property of natural language: the fact that word appearances are rare but *contagious* events – once we see a word, we are likely to see it again and again. Dirichlet is clearly superior to the unigram model where multiple occurrences of any word are completely independent of each other. However, all interactions in the model are limited to repetitions of the same word; unlike the mixture model, the Dirichlet cannot capture the fact that "carburetor" goes with "cars" but not with "politics". So we are not really justified in viewing the Dirichlet model as a mixture of an infinite number of topics; in fact the Dirichlet model is not capable of capturing more than a single topic.

4.2.4 Probabilistic Latent Semantic Indexing (pLSI)

The probabilistic LSI model was introduced by Hoffman in [58] and quickly gained acceptance in a number of text-modeling applications. In some circles pLSI is known as the *aspect model*, or as a *simplicial mixture* model. Conceptually, the roots of pLSI go back to methods of latent semantic analysis (LSA / LSI [40]), which involve singular-value decomposition of a matrix of real-valued observations. One of the main problems with applying LSA to language is that word observations are not real-valued. Natural text is a fundamentally discrete phenomenon, treating it as continuous observations is certainly possible, but constitutes a stretch of assumptions. pLSI was designed by Hoffman as a discrete counterpart of LSI and PCA/ICA techniques. Another way to look at pLSI is as an attempt to relax the assumptions made in the mixture model. Recall that a simple mixture model assumes that each string of text is generated by a *single* topic model T_z. This can be problematic for long strings of text, e.g. articles or editorials, which can discuss more than one subject. Hoffman's solution to the problem was simple and elegant: he allowed each string to contain a mixture of topics. Hypothetically, every word in a string could come from a different topic. Just like the mixture model, the pLSI model is defined by a set of k topic models $T_1 \ldots T_k$, and a mixture distribution $\pi(\cdot)$. The process of generating a given string $\mathbf{w} = w_1 \ldots w_n$ is as follows:

1. Pick a mixing distribution $\pi_{\mathbf{w}}$ corresponding to $\mathbf{w}$.
2. For each position in the string $i = 1 \ldots n$:
   a) Pick a topic $z \in \{1 \ldots k\}$ with probability $\pi_{\mathbf{w}}(z)$.
   b) Pick a word w_i from the topic z with probability $T_z(w_i)$

The overall likelihood of observing $\mathbf{w} = w_1 \ldots w_n$ under the above generative process is:

$$P_{lsi}(w_1 \ldots w_n) = \prod_{i=1}^{n} \left(\sum_{z=1}^{k} \pi_{\mathbf{w}}(z) T_z(w_i) \right) \tag{4.10}$$

When we compare the above equation to the mixture model (equation 4.3), two things strike the eye. First, the summation $\sum_z$ moved inside the product – this reflects the fact that topics are mixed inside the string on the level of a single word w_i. The second observation is worrisome: the mixture distribution $\pi_{\mathbf{w}}(\cdot)$ is conditioned on $\mathbf{w}$, the very string we are trying to generate. Technically, such parametrizations are not allowed in a generative model: it makes no sense to condition the probability of an event on itself. In a way, $\pi_{\mathbf{w}}$ represents an *oracle*, which tells us exactly in what proportion we should mix the topics for a particular string $\mathbf{w}$. Such distributions are not allowed because they usually do not add up to one over the space of all possible samples. A common way to turn equation (4.10) into a proper generative model, is to marginalize the model by summing over all strings $\mathbf{w}$ in the training set. A *generative pLSI* model assigns the following probability to a new, previously unobserved string $w_1 \ldots w_n$:

$$P_{lsi}(w_1 \ldots w_n) = \frac{1}{N} \sum_{\mathbf{w}} \prod_{i=1}^{n} \left(\sum_{z=1}^{k} \pi_{\mathbf{w}}(z) T_z(w_i) \right) \tag{4.11}$$

where the outside summation goes over every training string $\mathbf{w}$, and N represents the total number of training strings. When we discuss pLSI in the following sections, we will be referring to the generative version (equation 4.11), not the one based on equation (4.10).

The parameters of pLSI include the topic distributions $T_1 \ldots T_k$, and the mixture distributions $\pi_{\mathbf{w}}$ for every string $\mathbf{w}$ in the training collection $\mathcal{C}_{train}$. Parameters are estimated by assuming that the training strings represent an observation from equation (4.10). The goal is to find the topics $T_1 \ldots T_k$ and mixtures $\pi_{\mathbf{w}}$ that maximize the log-likelihood of $\mathcal{C}_{train}$:

$$\sum_{\mathbf{w} \in \mathcal{C}_{train}} \sum_{i=1}^{n_{\mathbf{w}}} \log \left(\sum_{z=1}^{k} \pi_{\mathbf{w}}(z) T_z(w_i) \right) \tag{4.12}$$

Maximizing of equation (4.12) has to be done iteratively. It is amenable to gradient methods, but a more common approach is to apply expectation maximization (EM); details can be found in [58]. Due to a very large number of mixing distributions $\pi_{\mathbf{w}}$, pLSI is extremely prone to over-fitting. Hoffmann [58] suggests using simulated annealing to temper the maximization process; other authors found it necessary to cluster the training strings, initialize the topic models T_z from clusters C_z, and then run only a handful of EM iterations to prevent over-fitting.

4.2.5 Latent Dirichlet Allocation

The main problem of pLSI comes from inappropriate generative semantics. Tying the mixing distribution $\pi_{\mathbf{w}}$ to individual strings $\mathbf{w}$ makes the model both confusing and susceptible to over-fitting. Blei, Ng and Jordan [12] introduced a new, semantically consistent aspect model, which attracted a tremendous amount of interest from the statistical learning community. They called the model *Latent Dirichlet Allocation*. The basic setup of the model closely resembles pLSI: the authors assume that the data can be explained by a set of k topics, each topic z is associated with a distribution T_z over our vocabulary, and each word in a given string can come from a mixture of these topics. The main contribution of LDA was in the way of defining the mixing distributions $\pi(z)$. Blei and colleagues make π a random variable, which means it can be any distribution over the topics $z = 1\ldots k$. Formally, any given mixture $\pi(\cdot)$ is a single point in the *topic simplex* $\mathbb{P}_k = \{x \in [0,1]^k : |x| = 1\}$, which includes all possible distributions over the numbers $1\ldots k$. To reflect the uncertainty about the specific value of $\pi(\cdot)$, the authors impose a density function $p_k(\pi)$ over the points $\pi \in \mathbb{P}_k$ in the simplex. Under these conditions, the process of generating a string $w_1\ldots w_n$ takes the following form:

1. Pick a mixing distribution $\pi(\cdot)$ from $\mathbb{P}_k$ with probability $p_k(\pi)$.
2. For each position in the string $i = 1\ldots n$:
   a) Pick a topic $z \in \{1\ldots k\}$ with probability $\pi(z)$.
   b) Pick a word w_i from the topic z with probability $T_z(w_i)$

The generative process behind LDA closely resembles pLSI, but has one important difference: instead of using an oracle $\pi_{\mathbf{w}}(\cdot)$ as the topic weighting distribution, the authors randomly sample $\pi(\cdot)$ from the density function $p_k(\pi)$. The likelihood assigned by LDA to a given string $w_1\ldots w_n$ is:

$$P_{lda}(w_1\ldots w_n) = \int_{\mathbb{P}_k} \left\{ \prod_{i=1}^{n} \sum_{z=1}^{k} \pi(z) T_z(w_i) \right\} p_k(\pi) d\pi \tag{4.13}$$

As their density function p_k, the authors pick the Dirichlet distribution (equation 4.7) with parameters $\alpha_1\ldots\alpha_k$:

$$p_k(\pi) = \Gamma\left(\sum_{z=1}^{k} \alpha_z\right) \prod_{z=1}^{k} \frac{\pi(z)^{\alpha_z - 1}}{\Gamma(\alpha_z)} \tag{4.14}$$

This is a natural choice since the topics T_z are multinomial distributions, and Dirichlet is a conjugate prior for the multinomial. Despite its complexity, the LDA model has a relatively small number of free parameters: k parameters for the Dirichlet density and $|\mathcal{V}|-1$ probabilities for each of the k topic models T_z. The total number of parameters is $k|\mathcal{V}|$, which is almost the same as $k|\mathcal{V}|-1$ for a simple mixture model with as many topics, and substantially smaller than $k(|\mathcal{V}|-1) + (k-1)|\mathcal{C}_{train}|$ for a similar pLSI model. The estimation of free parameters is done using variational inference methods; complete details are provided in [12, 13].

4.2.6 A brief summary

We have described five generative formalisms that can be used to model strings of text. We presented them in the order of increasing modeling power, and increasing complexity of parameter estimation. Two of these models (unigram and Dirichlet) can only model homogeneous collections of text, and do not offer much in terms of capturing possible dependencies between the words. The other three explicitly model a collection as a mixture of k topics. These models allow for various degrees of dependence between the words. More specifically, dependence between words a and b can be encoded by making both words prominent in some topic model T_z and not prominent in other topics. The LDA model in particular has proven to be quite effective in modeling dependencies in a wide array of applications.

4.2.7 Motivation for a new model

Any one of the above models could be adopted as a generative process associated with relevance. However, we are not going to do that. In our opinion all of these models make the same unreasonable assumption. All five models postulate that any collection of strings can be modeled by a finite set of k aspects or topics[1]. The assumption is easy to make because it is intuitively appealing. As humans, we like to categorize things, arrange them on shelves, assign labels to them. We are always lured towards generalization, tempted to say that this news story is about *sports* and that one talks about *politics*; this image is a *landscape* and that other one is a *portrait* mixed with some *abstract art.* Even the current book will in the end be categorized as something along the lines of [*computer science; information retrieval; formal models*]. Topics and topical hierarchies are appealing because they allow us to deal with the complexity of the surrounding world. However, there is absolutely no intrinsic reason for using them in generative models. The dependency structure of large collections rarely corresponds to any given set of topics or to any hierarchy. And then there is the eternal question of how many topics are appropriate for a given model. Over the years, researchers have proposed a wide array of formal and heuristic techniques for selecting k, but somehow the question never seems to go away. The reason is that it is incredibly hard even for humans to pick k, even if they are dealing with a very specific task and are highly experienced in this task. Ask a librarian how many Dewey decimal codes she uses in her section of the library, and you may get a specific number k. Then ask her if k is enough, if she ever wanted to add just one more for that odd set of books that didn't seem to fit well with everything else. Or, suppose we try to come up with a general set of categories for news reporting. This task has actually been carried out by the Linguistic Data Consortium as part of topic annotations for the TDT project [5]. The result is a set of 13 broad categories, ranging from *politics* to *scandals* to *natural disasters.* Among the thirteen, which category accounts for the largest number of reports? The largest category is called "miscellaneous", and it contains news that simply does not fit in any of the other categories. This is the reality of information. Topic labels and hierarchies give a terrific way to present information to a human user, but one should not expect magic from using them as a foundation for probabilistic models of real data.

Our generative model will not be based on any sort of latent topic or aspect structure. But before we introduce our ideas, we will summarize the existing

[1] $k = 1$ for the unigram and Dirichlet models.

models in terms of a common framework – the framework that will naturally lead to our own model.

4.3 A Common Framework for Generative Models

Recall that we started this chapter by saying that we are interested in modeling exchangeable sequences of words. According to de Finetti's theorem, the joint distribution of the variables will take the form of equation (3.1), which we repeat below for the reader's convenience:

$$P(w_1 \ldots w_n) = \int_{\Theta} \left\{ \prod_{i=1}^{n} P_{i,\theta}(w_i) \right\} p(d\theta) \tag{4.15}$$

where $P_{i,\theta}(w_i)$ is some distribution appropriate for dimension $\mathcal{S}_i$, θ denotes the set of parameters for that distribution, and $p(d\theta)$ is a probability measure over all possible settings of those parameters.

The above equation is very general, so it should come as no surprise that all of the models discussed in the previous section can be viewed as special cases. After all, we can define the individual probabilities $P_{i,\theta}(w_i)$ to be as elaborate as we want. However, in this section we will demonstrate that every model discussed in section 4.2 can be encompassed by a much simpler expression:

$$P(w_1 \ldots w_n) = \int_{\mathbb{P}_{\mathcal{V}}} \left\{ \prod_{i=1}^{n} u(w_i) \right\} p(du) \tag{4.16}$$

Equation (4.16) represents a very significant simplification from the general form (equation 4.15), and it is crucial to understand the difference. To arrive at equation (4.16), we have replaced the distribution $P_{i,\theta}(\cdot)$, which could represent an arbitrarily complex expression, with $u(\cdot)$, which is a simple unigram model. Also, instead of an abstract *parameter* space Θ, we now have the vocabulary *simplex* $\mathbb{P}_{\mathcal{V}} = \{u \in [0,1]^{\mathcal{V}} : |u| = 1\}$, which is the set of all possible unigram distributions. The crucial detail here is that we don't get to "define" u: it is just a point in $\mathbb{P}_{\mathcal{V}}$. The only component of equation (4.16) under our control is the $p(du)$. The generative process that corresponds to equation (4.16) would look like this:

1. Pick a unigram model u according to the *generative density* $p(du)$
2. For $i = 1 \ldots n$: pick word w_i from the unigram model u with probability $u(w_i)$

In the remainder of this section, we will show that all five models are equivalent to the above generative process. They only differ in the way they allocate the generative density $p(du)$ to the regions of $\mathbb{P}_{\mathcal{V}}$. The equivalence is fairly evident for the first three models, but it *should* come as a surprise that pLSI and LDA are also special cases of the formalism. In each of the five cases we will provide a brief geometric interpretation of the model. Geometrical details become particularly fascinating for the simplicial mixture models.

4.3.1 Unigram

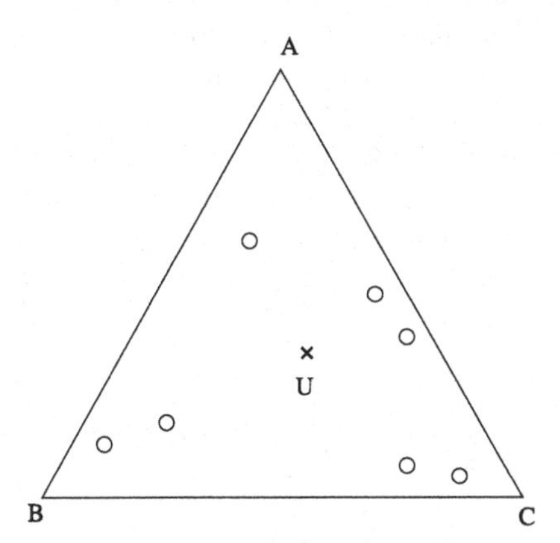

Fig. 4.1. A geometric view of the unigram model over a three-word vocabulary. The triangle represents the probability simplex. Circles reflect empirical distributions of words in training strings. The point U marks the place where the generative density is non-zero.

The case of a unigram model is particularly simple. In order to get equation (4.1), we force the generative density to be the Dirac delta function:

$$p_{uni}(\mathrm{d}u) = \begin{cases} 1 \text{ if } \mathrm{d}u{=}\{U\} \\ 0 \text{ otherwise} \end{cases} \tag{4.17}$$

The effect of equation (4.17) is that we place all the generative density on a single point $U \in \mathbb{P}_{\mathcal{V}}$. Strictly speaking, $p_{uni}(\mathrm{d}u)$ is not a *density* function, rather it is a probability *measure*. Accordingly, the integral in equation (4.16) is not the *Riemann* integral familiar to most engineers and scientists; it is the *Lebesgue* integral, which allows us to define and use all sorts of interesting probability measures. In the interest of accessibility to a wider audience, the remainder of this book will use the word *density* as a subsitute for *measure*. Similarly, all integrals from this point on will refer to *Lebesgue* integrals. A reader interested in the theory of Legesbue integration is invited to consult any graduate textbook on modern analysis.

The result of using the Dirac delta function as a generative density in equation (4.16) is as follows:

$$P(w_1 \ldots w_n) = \int_{\mathbb{P}_{\mathcal{V}}} \left\{ \prod_{i=1}^{n} u(w_i) \right\} p_{uni}(\mathrm{d}u) = \prod_{i=1}^{n} U(w_i) \tag{4.18}$$

Intuitively, the delta function simply picks out the necessary point $u = U$ and zeroes out the rest of the simplex $\mathbb{P}_{\mathcal{V}}$. We show a geometric view of the unigram model in Figure 4.1. We assume a toy vocabulary of three words $\{A, B, C\}$. The corresponding simplex is a triangle, each corner corresponds to a distribution that puts all the probability mass on a single word. The circles represent empirical distributions of words in training strings – the distributions we would get by counting how many times each word occurs in a given string, and then normalizing them so they add up to one. U marks our unigram urn – the point where all the generative density is concentrated on the simplex. The effect of using a unigram model is that we assume the single point U to be representative of all training strings (the circles).

4.3.2 Dirichlet

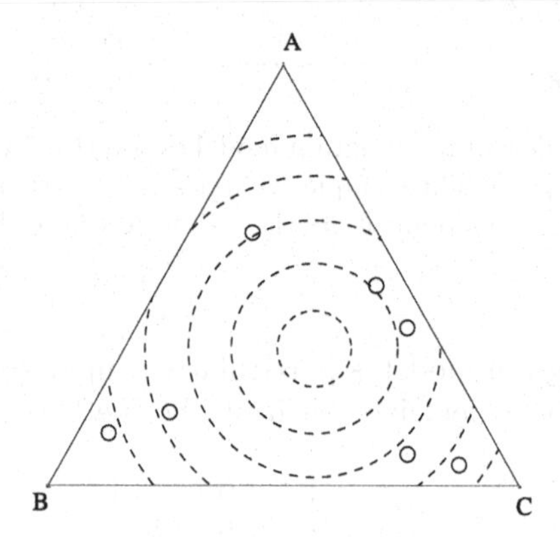

Fig. 4.2. Geometric view of the Dirichlet model. Circles reflect training strings. Dashed lines represent a Dirichlet density over the three-word probability simplex.

Showing the equivalence of a Dirichlet model is trivial: equations (4.16) and (4.6) are virtually identical. All we have to do is set the generative density $p_{dir}(\mathrm{d}u)$ to be the Dirichlet distribution with parameters α_v for $v \in \mathcal{V}$:

$$p_{dir}(\mathrm{d}u) = \mathrm{d}u \times \Gamma\left(\sum_{v \in \mathcal{V}} \alpha_v\right) \prod_{v \in \mathcal{V}} \frac{u(v)^{\alpha_v - 1}}{\Gamma(\alpha_v)} \tag{4.19}$$

We show a geometric view of the model in Figure 4.2. Everything is as before, only now instead of a single point U, the generative density is spread out over the entire simplex. Dashed circles represent possible isolines of the Dirichlet density. The mass would be concentrated around the center and would decrease towards the edges of the simplex.

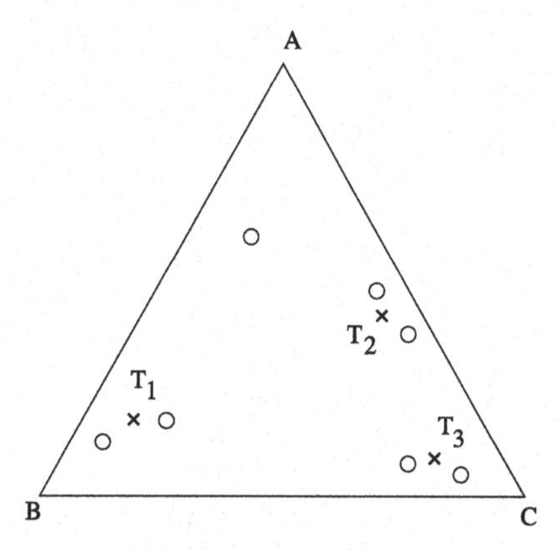

Fig. 4.3. Geometric view of the mixture model. Circles reflect training strings. Generative density is concentrated over the points T_1, T_2 and T_3, which correspond to three topics in the mixture model.

4.3.3 Mixture

The mixture model consists of a set of topics $T_1 \ldots T_k$, and a weighting function $\pi(z)$ for mixing them together. In order to get our equation (4.16) to look like the mixture model, we will define the generative density as follows:

$$p_{mix}(\mathrm{d}u) = \begin{cases} \pi(z) \text{ if } \mathrm{d}u = \{T_z\} \text{ for some topic } z \\ 0 \quad \text{otherwise} \end{cases} \tag{4.20}$$

Now if we plug equation (4.20) into (4.16), we will get exactly equation (4.3), which defines the mixture model. The effect of density $p_{mix}(\mathrm{d}u)$ is to pick out k points from the simplex, each point corresponding to one of the topic models T_z. The points are weighted by $\pi(z)$. The integral over the simplex turns into a summation over the k points of $\mathbb{P}_{\mathcal{V}}$. Geometrically, the effect of using a mixture model is shown in Figure 4.3. The simplex and the training strings (circles) are unchanged. The generative density is concentrated on the three points marked as T_1, T_2 and T_3. Each point represents one of the k=3 topics in the mixture model. The topics seem to provide a good representation for a majority of the training strings, but note what happens when we have an outlier. The uppermost circle in the simplex corresponds to a training string that doesn't look like any of the other strings. The mixture model will force this string to be associated with T_2, the closest topic model. Naturally, T_2 will not be a good representation of that string. As we will see, this is a recurring problem with aspect representations: whenever we assume a fixed number of topics, the model will do a good job of capturing the bulk of the data, but will completely misrepresent any outliers in the collection.

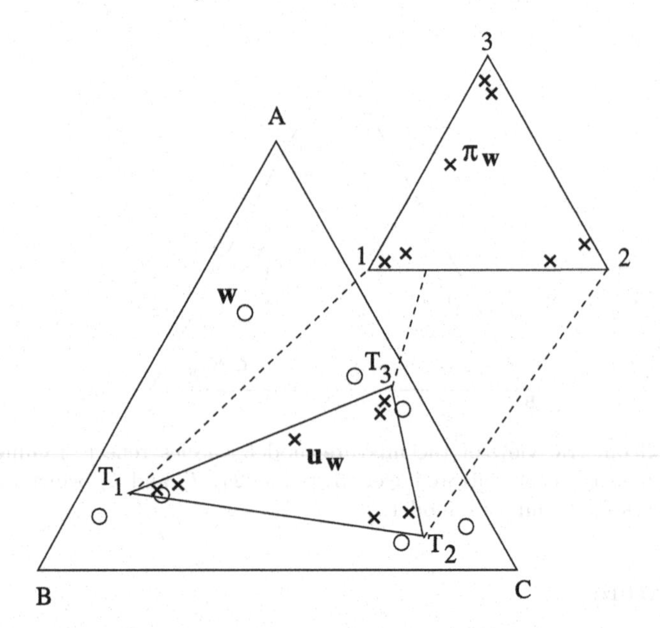

Fig. 4.4. Geometric view of the pLSI model. To every training string **w** the model assigns a topic mixing distribution $\pi_{\mathbf{w}}$. In turn, this distribution defines a single point $u_{\mathbf{w}}$ in the sub-simplex spanned by pLSI aspects T_1, T_2 and T_3. All the generative density in the model is concentrated over the points $u_{\mathbf{w}}$.

4.3.4 pLSI

The Probabilistic LSI model is substantially more complex than our previous cases, so we will make extensive use of geometry to illustrate our argument. Recall that pLSI, like the mixture model, consists of a set of k topics T_z which are mixed on a word level using the distribution $\pi_{\mathbf{w}}(z)$. Let us try to figure out what pLSI looks like geometrically. In Figure 4.4 we have our usual word simplex $\mathbb{P}_{\mathcal{V}}$ with words A, B and C in the corners. As before, the training strings **w** are represented by small circles. We also have another simplex, with corners marked by 1, 2 and 3, it represents the set of all possible ways to mix k=3 components. pLSI endows every training string **w** with its own mixing distribution $\pi_{\mathbf{w}}(\cdot)$. These distributions are points in that smaller[2] *topic simplex* $\mathbb{P}_k$. Each corner of $\mathbb{P}_k$ corresponds to a single pLSI topic. For example, if some $\pi_{\mathbf{w}}$ were placed in corner 1, that would mean that string **w** is best modeled by the first topic alone.

Now we come to the key part of our argument. We observe that for each point in $\mathbb{P}_k$ there exists a corresponding point in the word simplex $\mathbb{P}_{\mathcal{V}}$. The corners $1\ldots k$ of $\mathbb{P}_k$ correspond to the unigram topic models $T_1\ldots T_k$, which are

[2] In our example, there are three words in the vocabulary and three pLSI topics. In a realistic setting, the number of topics would be orders of magnitude smaller than vocabulary size.

the points in $\mathbb{P}_{\mathcal{V}}$. Naturally, any linear mixture of topics $1...k$ will correspond to a similar mixture of unigram models $T_1...T_k$. These mixtures will form a *sub-simplex* $\mathbb{P}_T$ with unigram models $T_1...T_k$ as the corners. The important consequence of the mapping is this: for any mixing distribution $\pi_{\mathbf{w}}(\cdot)$ over the topics, there will be one and only one corresponding point $u_{\mathbf{w}}$ in the simplex $\mathbb{P}_{\mathcal{V}}$. The conclusion we draw from this is somewhat startling:

A generative version of pLSI is equivalent to a regular mixture model.

Indeed, we can take equation (4.11), which defines the probability that generative pLSI would assign to a string $w_1...w_n$, and re-write it as follows:

$$P_{lsi}(w_1...w_n) = \frac{1}{N}\sum_{\mathbf{w}}\prod_{i=1}^{n}\sum_{z=1}^{k}\pi_{\mathbf{w}}(z)T_z(w_i)$$
$$= \frac{1}{N}\sum_{\mathbf{w}}\prod_{i=1}^{n}u_{\mathbf{w}}(w_i) \tag{4.21}$$

where $u_{\mathbf{w}}(\cdot)$ is a point in $\mathbb{P}_{\mathcal{V}}$ that corresponds to the mixing distribution $\pi_{\mathbf{w}}$:

$$u_{\mathbf{w}}(v) = \sum_{z=1}^{k}\pi_{\mathbf{w}}(z)T_z(v) \tag{4.22}$$

Why is this important? Because it gives us a much more realistic view of what pLSI really is. The universal view of pLSI is that it is a *simplicial*, or word-level, mixture with k aspects. We have just demonstrated that pLSI is a regular, string-level mixture with N components, one for each training string.[3] Another interesting observation is that all N components are restricted to the sub-simplex $\mathbb{P}_T$, formed by the original aspects. In a sense, pLSI defines a low-dimensional sandbox within the original parameter space $\mathbb{P}_{\mathcal{V}}$, and then picks N points from that sandbox to represent the strings in a collection. Since pLSI really has N mixture components, should we expect it to perform as a general, unrestricted mixture model with N topics? The answer is a resounding no: pLSI components are restricted by the k-aspect sandbox, they cannot travel to the regions of parameter space that contain outliers. pLSI will be able to capture more than a k-topic mixture, but nowhere near the unrestricted N-topic mixture.

Since pLSI turns out to be a restricted mixture, it is quite easy to define the generative density $p(du)$ that will make equation (4.16) equivalent to pLSI:

$$p_{lsi}(du) = \begin{cases} \frac{1}{N} & \text{if } du = \{u_{\mathbf{w}}\} \text{ for some training string } \mathbf{w} \\ 0 & \text{otherwise} \end{cases} \tag{4.23}$$

[3] Our assertion refers to the marginalized (generative) version of pLSI. The original version of pLSI [58] is equivalent to a regular mixture with *only one* component, selected by an "oracle" mixture. In this way, the original pLSI is equivalent to the simple unigram model, it just gets to pick a perfect unigram urn for each new string.

Here $u_{\mathbf{w}}$ is defined according to equation (4.22). It is easy to verify that plugging $p_{lsi}(du)$ into equation (4.16) will give us the pLSI equation (4.11).

4.3.5 LDA

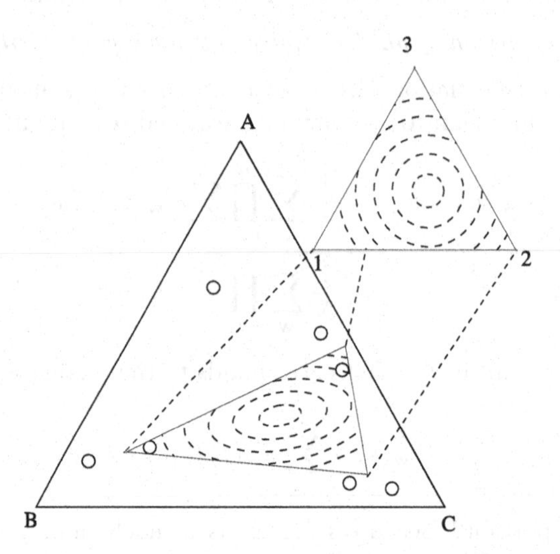

Fig. 4.5. Geometric view of the LDA model. The generative density is restricted to the sub-simplex spanned by the k=3 LDA topics T_z inside the word simplex $A{-}B{-}C$. The density is induced by the Dirichlet prior over the mixing distributions $\pi(z)$, which are points in the smaller simplex $1{-}2{-}3$.

Our final case concerns the LDA model. The derivation is perhaps the hardest, but it will be greatly simplified by the results we proved in the previous section. Recall that LDA consists of a set of k topic models $\{T_z : z{=}1\ldots k\}$, which are mixed on the word level using the mixing distribution $\pi(z)$. What makes LDA different from pLSI is that the mixing distribution π is not fixed but stochastically drawn from the simplex $\mathbb{P}_k$ under a Dirichlet prior $p_k(\pi)$. In Figure 4.5 we try to visualize LDA geometrically. The small triangle with corners 1, 2 and 3 represents the *topic simplex* $\mathbb{P}_k$. It is the set of all possible mixtures of k=3 components. Mixing distributions $\pi(\cdot)$ are elements of that set, and the dashed lines indicate a Dirichlet density $p_k(\cdot)$ over the set. The corners of $\mathbb{P}_k$ correspond to the topic models T_z, which are points in the vocabulary simplex $\mathbb{P}_{\mathcal{V}}$, the big triangle with corners A, B and C. The topic models $T_1\ldots T_k$ make up the corners of a *sub-simplex* $\mathbb{P}_T$, a smaller triangle inside $\mathbb{P}_{\mathcal{V}}$. Every point u_π in the sub-simplex $\mathbb{P}_T$ corresponds to a mixing distribution $\pi(\cdot)$ in $\mathbb{P}_k$. Specifically, the point u_π is the result of mixing the unigram topic models $T_1\ldots T_k$, weighted by $\pi(\cdot)$:

$$u_\pi(v) = \sum_{z=1}^{k} \pi(z) T_z(v) \tag{4.24}$$

The key observation to make is that Dirichlet density $p_k(\pi)$ over possible mixtures $\pi(\cdot)$ automatically induces a similar density $p_{lda}(u_\pi)$ over the unigram models u_π. This density is shown as dashed lines inside the sub-simplex $\mathbb{P}_k$. Assuming that topic models $T_1 \ldots T_k$ are not redundant[4], this new density can be expressed very simply:

$$p_{lda}(u_\pi) = p_k(\pi) \tag{4.25}$$

This density is defined for every point u_π in the sub-simplex $\mathbb{P}_T$. There are no holes because every point u_π corresponds to a unique mixing distribution $\pi \in \mathbb{P}_k$. However, the density $p_{lda}(u_\pi)$ is not defined for all points u in $\mathbb{P}_\mathcal{V}$, since some points in the larger simplex will not have a pre-image in $\mathbb{P}_k$. Fortunately, this is quite easy to fix. Since $\mathbb{P}_T$ is completely embedded inside $\mathbb{P}_\mathcal{V}$, we simply assign zero mass to any point $u \in \mathbb{P}_\mathcal{V}$ which happens to lie outside the sub-simplex $\mathbb{P}_T$. This gives us the following generative density:

$$p_{lda}(du) = \begin{cases} \frac{p_k(\pi)}{V(\mathbb{P}_T)} \times du & \text{if } u = u_\pi \text{ for some mixing distribution } \pi(\cdot) \\ 0 & \text{otherwise} \end{cases} \tag{4.26}$$

Here u_π is given by equation (4.24), and $p_k(\pi)$ is the Dirichlet density over mixtures, defined in equation (4.14). $V(\mathbb{P}_T)$ represents the *volume* of the sub-simplex $\mathbb{P}_T$, dividing by it ensures that the generative density $p_{lda}(du)$ will integrate to one over the whole simplex $\mathbb{P}_\mathcal{V}$.

Let $f : \mathbb{P}_k \mapsto \mathbb{P}_T$ denote the function that takes a mixture distribution $\pi(\cdot)$ and produces the corresponding unigram model u_π according to equation (4.24). Since the topics $T_1 \ldots T_k$ are not redundant, f is a one-to-one mapping. Consequently, the inverse f^{-1} is well-defined: it takes a unigram model $u \in \mathbb{P}_T$ and maps it back to the mixture $\pi(\cdot)$ that produced u. Now let's take the generative density $p_{lda}(du)$, plug it into equation (4.16) and verify that we recover the LDA model:

$$\begin{aligned} P(w_1 \ldots w_n) &= \int_{\mathbb{P}_\mathcal{V}} \left\{ \prod_{i=1}^{n} u(w_i) \right\} p_{lda}(du) \\ &= \int_{f(\mathbb{P}_k)} \left\{ \prod_{i=1}^{n} f(f^{-1}(u))(w_i) \right\} \frac{p_k(f^{-1}(u))}{V(\mathbb{P}_T)} du \end{aligned}$$

[4] Topic models $T_1 \ldots T_k$ are redundant if they form a polytope of dimension smaller than $k-1$. Equivalently, topic models are redundant if two different mixing distributions $\pi_1 \neq \pi_2$ can produce the same mixture model: $\sum_z \pi_1(z) T_z = \sum_z \pi_2(z) T_z$. This is theoretically possible, but almost never happens in practice because k is orders of magnitude smaller than the dimensionality of $\mathbb{P}_\mathcal{V}$ (number of words in the vocabulary).

$$= \int_{\mathbb{P}_k} \left\{ \prod_{i=1}^{n} f(\pi)(w_i) \right\} p_k(\pi) d\pi$$

$$= \int_{\mathbb{P}_k} \left\{ \prod_{i=1}^{n} \sum_{z=1}^{k} \pi(z) T_z(w_i) \right\} p_k(\pi) d\pi \qquad (4.27)$$

Equation (4.27) leads us to the following conclusion:

LDA is equivalent to a Dirichlet model restricted to the topic sub-simplex.

The conclusion is in line with our result on pLSI: it appears that both models are equivalent to regular mixtures. There appears to be nothing special about the fact that these models mix the topics on a word level – it is equivalent to string-level mixing. Furthermore, both pLSI and LDA restrict the generative density to the same $k-1$ dimensional sandbox bounded by the aspects $T_1 \ldots T_k$. The main difference is that pLSI concentrates the density of N discrete points, while LDA spreads it out over the entire sandbox.

4.3.6 A note on graphical models

In the previous sections we demonstrated that equation (4.16) captures the following five generative models: unigram, Dirichlet, mixture, Probabilistic Latent Semantic Indexing and Latent Dirichlet Allocation. These five models represent the most dominant frameworks for dealing with exchangeable sequences. The last two models: pLSI, and especially LDA are widely regarded to be the most effective ways for capturing topical content, precisely because they allow aspects to be mixed at the word level of a string.

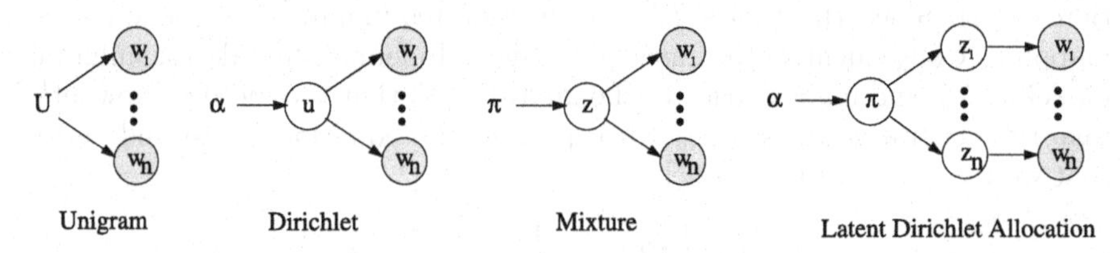

Fig. 4.6. Graphical dependence networks corresponding to unigram, Dirichlet, Mixture model and LDA. Observable variables (words w_i) are shaded. Unshaded circles represent hidden variables. Labels without circles reflect priors. Arrows mark dependencies.

On the surface, all five models look very different from each other. For example, LDA involves many more variables than a simple unigram model, and dependencies between these variables are carefully crafted into a very specific structure. As an illustration, Figure 4.6 shows the graphical diagrams corresponding of the unigram model, the Dirichlet model, the Mixture model and the Latent Dirichlet Allocation model. The graphical networks are intuitive, they give a clear picture of dependencies in each model, and are well-suited for constructing models that are easy to interpret. For example, in the LDA network we see that there is a set of topics (z), and we are free to pick a different topic for every word w_i.

While we certainly do not want to argue about the usefullness of graphical models, we would like to point out that they are somewhat misleading. Figure 4.6 alone does not make it obvious that the Dirichlet model cannot represent multiple topics. In fact the Dirichlet graphical network looks *exactly* like the mixture model network, even though the two models have very different behavior. Similarly, by looking at the networks we cannot see that the only difference between LDA and the Dirichlet is that the former restricts generative density to the topic sub-simplex.

In our opinion, the geometric analysis of the models is much more illuminating. It allowed us to place the different models into the same framework and focus on what really makes the models behave differently: the way they allocate generative density to the unigram points u. We believe that focusing *directly* on density allocation will ultimately lead to more powerful predictive models. We admit that these new models may not be nearly as intuitive as topic-based models such as pLSI and LDA, but as we mentioned before, we will opt for predictive power over elegance.

4.4 Kernel-based Allocation of Generative Density

We will now build upon the main result of the previous section. As we demonstrated, a large class of generative models can be represented by equation (4.16), which we re-state below for reading convenience:

$$P(w_1 \ldots w_n) = \int_{\mathbb{P}_{\mathcal{V}}} \left\{ \prod_{i=1}^{n} u(w_i) \right\} p(du)$$

The main difference between the models is not their dependence structure, but the way in which they allocate the density $p(du)$ to various regions of $\mathbb{P}_{\mathcal{V}}$. In this section we will construct two generative models by suggesting two possible choices for $p(du)$. We will not discuss the graphical structure of our models, focusing instead on their geometric representations, which we believe to be more expressive.

We have a collection of training strings $\mathcal{C}_{train}$, and would like to construct a generative model for it. We know that our generative model will take the form of equation (4.16), so instead of using maximum likelihood or Bayesian approaches, we will induce the model by *directly* constructing the density function $p(du)$. There exist a wide number of approaches for constructing a density function from a training collection. When we are faced with high-dimensional spaces and a small number of training examples[5], one of the best density estimation techniques is the *method of kernels*. An excellent exposition of the method can be found in [125]. The idea is to associate a *kernel* $K_{\mathbf{w}}(\cdot)$ with every training point $\mathbf{w}$. A kernel is a non-negative function, it is usually symmetric about zero, and must integrate to 1 over the space for which we are constructing our density. A kernel-based density estimate takes the form:

$$p_{ker}(du) = \frac{1}{N}\sum_{\mathbf{w}} K_{\mathbf{w}}(du) \tag{4.28}$$

The summation is performed over all $\mathbf{w} \in \mathcal{C}_{train}$. In a sense, a kernel density estimate is a mixture of a bunch of mini-densities, one associated with each training point $\mathbf{w}$. Kernel estimates are surprisingly powerful, especially in the face of limited examples. Perhaps most importantly, they have no tendency to ignore the outliers. Since each training point $\mathbf{w}$ has a kernel dedicated entirely to it, we are guaranteed that every outlier will have some probability mass floating around it. In the following two sections we will propose two possible ways of selecting a kernel $K_{\mathbf{w}}$ for training strings $\mathbf{w}$.

4.4.1 Delta kernel

Recall that whenever we presented a geometric interpretation of a generative model, we always had a set of small circles scattered over the word simplex $\mathbb{P}_{\mathcal{V}}$. Each of these circles corresponds to an empirical distribution $u_{\mathbf{w}}(\cdot)$ of words in that string. To get $u_{\mathbf{w}}(v)$ we simply count how many times each word v occurred in the string $\mathbf{w}$, and then normalize the counts to sum to one. The main problem with empirical distributions is that they contain many zeros. The vast majority of words will not occur in any given string. For example, a typical document in a TREC collection contains no more than a thousand distinct words, whereas the total vocabulary size is 2 or 3 orders of magnitude larger. This means that empirical distributions $u_{\mathbf{w}}$ will concentrate around the edges and corners of the generative simplex $\mathbb{P}_{\mathcal{V}}$. Points on the edges turn out to be somewhat problematic, so we will assume that empirical distributions $u_{\mathbf{w}}(\cdot)$ are smoothed out with background word probabilities $U(w)$:

$$u_{\mathbf{w}}(v) = \lambda\frac{n(v,\mathbf{w})}{|\mathbf{w}|} + (1-\lambda)U(v) \tag{4.29}$$

The background probabilities $U(v)$ come directly from equation (4.1), with λ as the parameter that controls the degree of smoothing. The effect is to pull the points $u_{\mathbf{w}}$ from the edges of the simplex towards the point U in the middle. Now we can define our first kernel $K_{\mathbf{w}}$:

[5] In many text collections, number of dimensions (unique words) is larger than the number of documents.

$$K_{\delta,\mathbf{w}}(\mathrm{d}u) = \begin{cases} 1 \text{ if } \mathrm{d}u = \{u_{\mathbf{w}}\} \\ 0 \text{ otherwise} \end{cases} \tag{4.30}$$

The kernel $K_{\delta,\mathbf{w}}$ is a delta function – it places all its probability mass on a single point $u_{\mathbf{w}}$. The corresponding density estimate $p_{dir}(\mathrm{d}u)$ is a set of N spikes, one for each training string $\mathbf{w}$. Each spike carries equal probability mass $\frac{1}{N}$, where N is the number of training strings. Figure 4.7 gives a geometric view of the model. The details are simple: the generative mass, denoted by crosses, will be concentrated on a set of N points, each corresponding to an empirical distribution $u_{\mathbf{w}}$ of some training string $\mathbf{w}$. A generative model based on Figure 4.7 will assign the following probability to a new string $w_1 \ldots w_n$:

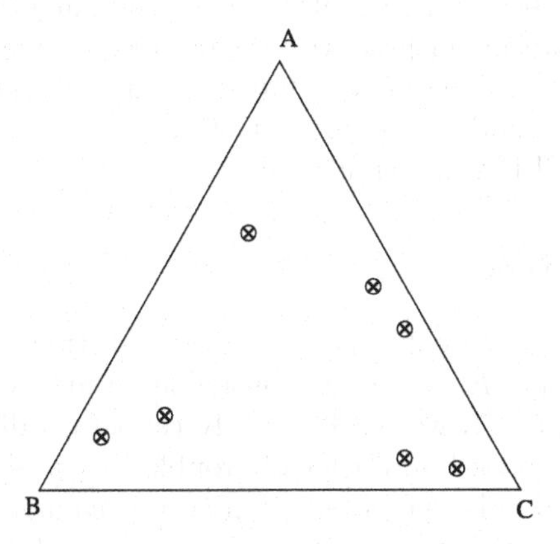

Fig. 4.7. Density allocation with delta-kernels. The generative mass is concentrated over N points corresponding to empirical distributions of the N training strings.

$$\begin{aligned} P(w_1 \ldots w_n) &= \int_{\mathbb{P}_{\mathcal{V}}} \left\{ \prod_{i=1}^{n} u(w_i) \right\} \left\{ \frac{1}{N} \sum_{\mathbf{w}} K_{\delta,\mathbf{w}}(\mathrm{d}u) \right\} \\ &= \frac{1}{N} \sum_{\mathbf{w}} \int_{\mathbb{P}_{\mathcal{V}}} \left\{ \prod_{i=1}^{n} u(w_i) \right\} K_{\delta,\mathbf{w}}(\mathrm{d}u) \\ &= \frac{1}{N} \sum_{\mathbf{w}} \prod_{i=1}^{n} u_{\mathbf{w}}(w_i) \end{aligned} \tag{4.31}$$

Keep in mind that since section (4.3.1) we have been using the word *density* to refer to *measures*, and that the integral $\int_{\mathbb{P}_{\mathcal{V}}}$ is the *Lebesgue* integral.

Training the model

One should immediately recognize equation (4.31) as a special case of the mixture model with N components (eqiation 4.3). However, there is one very important difference that should not go unmentioned. A general mixture model with N topics has $N\times|\mathcal{V}|$ free parameters: we must estimate $|\mathcal{V}|-1$ word probabilities for each of the N topic models, and then $N-1$ additional parameters for the mixing distribution $\pi(z)$. The model given by equation (4.31) has *one* free parameter: the constant λ, which determines the degree of smoothing for empirical distributions $u_{\mathbf{w}}$. There are no other free parameters, since the empirical distributions $u_{\mathbf{w}}(v)$ are unambiguously specified by the training set. As a result, training the model is extremely easy and fast: all we have to do is find the value of λ that optimizes our favorite criterion on a held-out portion of the training set. One may counter that we have simply moved computational complexity from the training to the testing phase of the model. There is a grain of truth to that: once we have the value of λ, computing equation (4.31) is expensive, since it involves a summation over every training example $\mathbf{w}$. However, in practice we can apply a number of algorithmic optimizations that make the model quite fast. In our experience, the overall time required by the model (training + testing) is a small fraction of what is required by pLSI or LDA in a similar setting.

4.4.2 Dirichlet kernel

The delta kernel $K_{\delta,\mathbf{w}}$ presented in the previous section is a rather odd kind of kernel. In fact, using $K_{\delta,\mathbf{w}}$ is only reasonable because we applied it on top of a generative simplex. If we tried to apply the kernel directly to the space of observations we would be in a lot of trouble: by its very definition $K_{\delta,\mathbf{w}}$ can generate one and only one observation ($\mathbf{w}$), it cannot possibly generalize to new observations. In our case, the generalization ability was ensured by applying $K_{\delta,\mathbf{w}}$ to the space of distributions $\mathbb{P}_{\mathcal{V}}$. Every point $u(\cdot)$ in the interior of $\mathbb{P}_{\mathcal{V}}$ can generate any observation string $\mathbf{w}$, because it assigns a non-zero probability $u(v)$ to every word v. Zero probabilities happen only on the edges, which is precisely why we use equation (4.29) to pull our kernels away from the edges of the simplex. The problem with $K_{\delta,\mathbf{w}}$ is not its ability to generalize, but rather the relatively high variance of the resulting density estimator. Recall that with the delta kernel, our density estimate $p_{ker}(du)$ is a set of spikes, one for each training example. If we take a different training set $\mathcal{C}_{train}$, we will end up with a completely different set of spikes. One way to reduce the variance is to have each spike spread the mass in some region around itself – effectively turning spikes into a set of bumps. Geometrically, the resulting generative model would look like Figure 4.8. Each training example would generate a density around itself, and these densities would overlap to produce some kind of a bumpy surface over the simplex A–B–C. This surface will act as the generative density $p_{ker}(du)$. The variance with respect to different training sets will be much lower than for a set of spikes. We can achieve the geometry of Figure 4.8 by replacing the delta kernel $K_{\delta,\mathbf{w}}$ with a kernel based on the Dirichlet distribution:

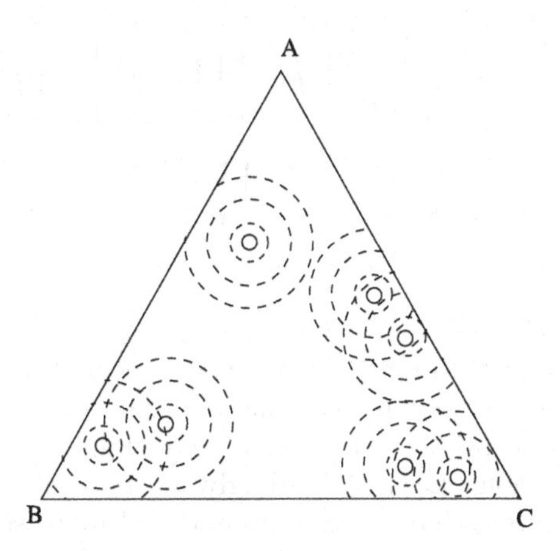

Fig. 4.8. Density allocation with Dirichlet kernels. Each training example induces a Dirichlet bump in the simplex $A-B-C$. The bumps (shown with dashed lines) overlap to create some overall generative density $p(du)$ over the simplex.

$$K_{\alpha,\mathbf{w}}(du) = du \times \Gamma\left(\sum_{v\in\mathcal{V}} \alpha_{\mathbf{w}}(w)\right) \prod_{v\in\mathcal{V}} \frac{u(v)^{\alpha_{\mathbf{w}}(v)-1}}{\Gamma(\alpha_{\mathbf{w}}(v))} \tag{4.32}$$

We chose the Dirichlet distribution because it is a conjugate prior for the unigram models $u(\cdot)$. The position and width of each Dirichlet bump is determined by a set of parameters $\alpha_{\mathbf{w}}(v)$, one for each word in the vocabulary. The parameters can be set in a number of ways, for instance:

$$\alpha_{\mathbf{w}}(v) \quad = \quad \lambda \cdot n(v, \mathbf{w}) \;+\; \sum_{\mathbf{x}\in\mathcal{C}_{train}} n(v, \mathbf{x}) \;+\; 1 \tag{4.33}$$

Here $n(v, \mathbf{w})$ gives the number of times we observe word v in string $\mathbf{w}$. The summation goes over all training strings $\mathbf{x}$, and λ is a constant that determines whether resulting Dirichlet bumps will be flat or spiky. Equation (4.33) looks like a heuristic, but in fact it can be derived as follows. We initialize the parameters by setting $\alpha_{\mathbf{w}}(v) = 1$, which gives us a uniform prior. Then we observe the entire training set $\mathcal{C}_{train}$. Next we observe λ copies of the string $\mathbf{w}$. The posterior expectation of $\alpha_{\mathbf{w}}(v)$ conditioned on these observations is exactly what we see in equation (4.33).

The benefit of using Dirichlet kernels comes to light when we try to express the overall likelihood of observing a new string $w_1 \ldots w_n$ from a generative model based on $K_{\alpha,\mathbf{w}}(du)$:

$$P(w_1 \ldots w_n) = \int_{\mathbb{P}_{\mathcal{V}}} \left\{ \prod_{i=1}^{n} u(w_i) \right\} \left\{ \frac{1}{N} \sum_{\mathbf{w}} K_{\alpha,\mathbf{w}}(du) \right\}$$

$$
\begin{aligned}
&= \frac{1}{N} \sum_{\mathbf{w}} \frac{\Gamma(\sum_v \alpha_{\mathbf{w}}(v))}{\prod_v \Gamma(\alpha_{\mathbf{w}}(v))} \int_{\mathbb{P}_{\mathcal{V}}} \left\{ \prod_{i=1}^{n} u(w_i) \right\} \left\{ \prod_{v \in \mathcal{V}} u(v)^{\alpha_{\mathbf{w}}(v)-1} \right\} du \\
&= \frac{1}{N} \sum_{\mathbf{w}} \frac{\Gamma(\sum_v \alpha_{\mathbf{w}}(v))}{\Gamma(\sum_v \alpha_{\mathbf{w}}(v) + n)} \prod_{v \in \mathcal{V}} \frac{\Gamma(\alpha_{\mathbf{w}}(v) + n(v, w_1 \ldots w_n))}{\Gamma(\alpha_{\mathbf{w}}(v))}
\end{aligned}
\tag{4.34}
$$

The good news stemming from equation (4.34) is that we do not have to perform numerical integration: $\int_{\mathbb{P}_{\mathcal{V}}}$ has a closed form solution solely because Dirichlet is a conjugate prior to the unigram. The bad news is that equation (4.34) is not susceptible to the same algorithmic trickery that allowed us to speed up equation (4.31). Accordingly, we have not been able to test equation (4.34) in large-scale retrieval scenarios. But in section 4.5 we will demonstrate that it constitutes an effective generative model, giving us a strong motivation for developing algorithms that will make equation (4.34) feasible.

Training the model

We observe that equation (4.34) involves just one free parameter: the constant λ which acts as inverse *bandwidth* of a kernel. Large values of λ yield kernels that are highly peaked, low values flatten out the bumps and also move them towards the center of the simplex. We can find the appropriate value of λ by performing a simple brute-force search for the value that leads to highest likelihood for a held-out portion of the training set. Using a held-out subset is absolutely critical: if we do not, the model will quickly overfit, by driving λ towards positive infinity. This should come as no surprise: after all we are dealing with a kernel-based method. Curiously, very large values of λ will collapse the Dirichlet bumps into the delta spikes we discussed in the previous section. It will also drive the spikes towards the edges of the simplex $\mathbb{P}_{\mathcal{V}}$.

4.4.3 Advantages of kernel-based allocation

In the current section we proposed a kernel-based approach to distributing the probability mass over the generative simplex $\mathbb{P}_{\mathcal{V}}$. We believe this approach leads to novel generative models that have a number of distinct advantages over existing formulations. In this section we will briefly describe four of these advantages.

Handling rare events

Aspect-based models (pLSI, LDA) do an excellent job of capturing the bulk of the data in any collection. However, they are poorly suited for modeling outliers and rare events. Any outlier will automatically be folded into the topic structure, assigned to the closest aspect or mixture of aspects, even if these aspects really have nothing to do with the outlier string. Researchers have reported that aspect-based models show excellent performance on a large number of tasks, but most of these tasks are not affected by outlier events. These are tasks such as predictive modeling, clustering or coarse classification, where it is imperative to assign reasonable likelihoods to the majority of strings. What happens to outliers really does not affect performance on these tasks: outlying events are so rare they have no chance of influencing the objective function.

However, there are other tasks where rare events are extremely important. In information retrieval, the user may be searching for a very specific subject, discussed by a very small set of documents. Or he may be interested in finding a rare piece of video footage. In Topic Detection and Tracking some of the most interesting events are the ones that receive a very small amount of news coverage. And for large TDT events it is essential to identify them as quickly as possible, when there are only one or two early reports on them. Aspect models are not suitable for these environments because they will tend to represent rare events as the mixture of prominent events. A kernel-based approach is much more appealing in this case. The outliers are not forced into a larger structure. Every rare event is guaranteed to have some small amount of probability mass around it.

The above argument does not mean that a kernel-based approach will only work for rare events. On the contrary, if a large number of examples fall into some region of $\mathbb{P}_{\mathcal{V}}$, their kernels will overlap and result in higher generative density over that region. For large events a kernel-based approach will behave in a way similar to the aspect model.

No structural assumptions

At its core, any topic model is a *dimensionality reduction* technique. The basic assumption in these models is that our observations are not as high-dimensional as they look. Usually we take every word to be a separate dimension, but words are not independent, they correlate in strange ways. Because of these correlations, we should not expect our collection to have a $|\mathcal{V}|$-dimensional spherical structure. It is more likely to be some peculiar manifold – a shape with intrinsic dimensionality much lower than $|\mathcal{V}|$. An aspect model like pLSI or LDA will try to "hug" this shape, approximate it by placing topics T_z as its vertices. In doing this, the aspect model will significantly reduce the dimensionality, without hurting the structure of the data. A smaller number of dimensions is very beneficial in a model because it means less sparseness, better estimates and less redundancy among the remaining dimensions.

The above reasoning is perfectly valid. We believe it is very plausible that the dimensionality of a text collection is not $\mathcal{V}$, that the data is embedded in some low-dimensional shape. But let us consider how mixture models approximate that shape. Any aspect model will approximate the shape by a *linear* polytope with topics placed in the corners. That may be perfectly correct, but it does constitute a strong assumption about the structure of our data. In other words, the shape may be low-dimensional, but why do we assume it is linear, or even convex? An approach based on kernels also makes an attempt to "hug" the shape enclosing the data, but it does the hugging in a very different manner. In a way it places a small sphere around each data point, and allows the sphere to touch and fuse with the neighboring spheres. The result is something like a thick ribbon winding through space. If we ignore the thickness, we get a low-dimensional *manifold* that tracks the data points. But this manifold does not have to be a polytope, as in the aspect model: there is no assumption of linear structure.

Easy to train

Kernel-based approaches are essentially *non-parametric*. As we described before, our methods have only one free parameter λ that controls the degree of estimator variance. One should not take this to mean that our model constitutes a compact representation with one parameter to store. Quite the opposite, a kernel-based model involves literally billions of parameters: we can easily have 100,000 training strings $\mathbf{w}$, over a vocabulary with over 100,000 words v. This would give us 10,000,000,000 Dirichlet parameters $\alpha_{\mathbf{w},v}$. But the important thing to realize is that these parameters are not free, they are completely determined by the data. We do not have to learn any of these parameters, we just need to create the infrastructure which allows us to store and access them efficiently. Once we have that infrastructure, "training" the model involves learning just one parameter λ, which is trivial. To contrast that, aspect models are fully parametric and routinely require inference for hundreds of thousands of parameters, e.g. 100 aspects with vocabulary trimmed to 1000 or so words.

Allows discriminative training

We would like to stress another important consequence of using a non-parametric approach. Even though we are talking about a generative model, the objective function we use for training λ does not have to be limited to likelihood. Since we have only one parameter the simple brute search will be just as fast as any other search method. This means that we don't have to stick to "nice" objective functions that facilitate gradient methods or EM. Instead, we can directly maximize the criterion that is appropriate for the problem at hand, regardless of whether that criterion is convex, differentiable or even continuous. Likelihood is the appropriate objective when we are specifically interested in the predictive qualities of a model. For ad-hoc retrieval we should maximize the average precision. For topic detection we should minimize the TDT cost function. To summarize: while our model is *generative*, it allows for a healthy bit of *discriminative* tuning.

4.5 Predictive Effectiveness of Kernel-based Allocation

We do not want to conclude this section without providing some empirical support for the alleged advantages of the proposed approach. In the present section we will briefly discuss a small scale experiment showing that kernel-based allocation does indeed lead to an effective generative model.

Consider a simple task of predictive text modeling. We have a collection of exchangeable strings $\mathcal{C}$. The collection will be partitioned into two subsets: the training set $\mathcal{A}$ and the testing set $\mathcal{B}$. Our goal is to use the training strings to estimate an effective generative model $P_{\mathcal{A}}(\mathbf{w})$. As we discussed in section 4.1.1, a model is effective if it does a good job of predicting the unseen testing portion $\mathcal{B}$ of our collection. *Predictive perplexity* is a widely accepted measure of predictive accuracy of a model; the measure is defined as:

$$\text{Perplexity}(\mathcal{B};\mathcal{A}) = \exp\left\{-\frac{1}{N_B}\sum_{\mathbf{w}\in\mathcal{B}} \log P_{\mathcal{A}}(\mathbf{w})\right\} \tag{4.35}$$

Here $P_{\mathcal{A}}(\mathbf{w})$ is the likelihood that our generative model will assign to a previously unseen string $\mathbf{w}$. The quantity inside the exponent is called the *entropy* of the data. The summation goes over all testing strings $\mathbf{w}$, and $N_{\mathcal{B}}$ represents the combined length of all testing strings. The logarithm typically has base 2, which enables interpreting the entropy in terms of *bits* of information.

Predictive perplexity has a very intuitive interpretation. A perplexity of K means that for each testing word the model is as *perplexed* as if it were choosing between K equally attractive options. For example, a perplexity of 1 means that there is no uncertainty at all, the model is absolutely sure of which word is going to come next. A perplexity of 2 means that the system is as uncertain as if it ruled out all but two words, but the remaining two were equally plausible. We would like to stress that with perplexity **lower means better**.

We will compare predictive perplexity of the following four models. **(i)** the unigram model (section 4.2.1), **(ii)** the Dirichlet model (section 4.2.3), **(iii)** density allocation with delta kernels (section 4.4.1) and **(iv)** density allocation with Dirichlet kernels (section 4.4.1). Our experiments will be done on two datasets. The first dataset contains 200 documents from the 1998 Associated Press corpus [83]. The documents contain 11,334 distinct words; average document length is 460. The second dataset contains 200 documents from the TDT-1 corpus [24]. TDT documents are slightly longer, average length is 540 words, but the number of distinct words is somewhat smaller: 9,379. In both cases we used the odd documents for estimating the model and even documents for testing.

The results are shown in Figure 4.9. We show testing-set perplexity as a function of the smoothing parameter λ, which is the only free parameter for all four models. The first observation we make from Figure 4.9 is that both kernel-based approaches lead to a dramatic reduction is predictive perplexity

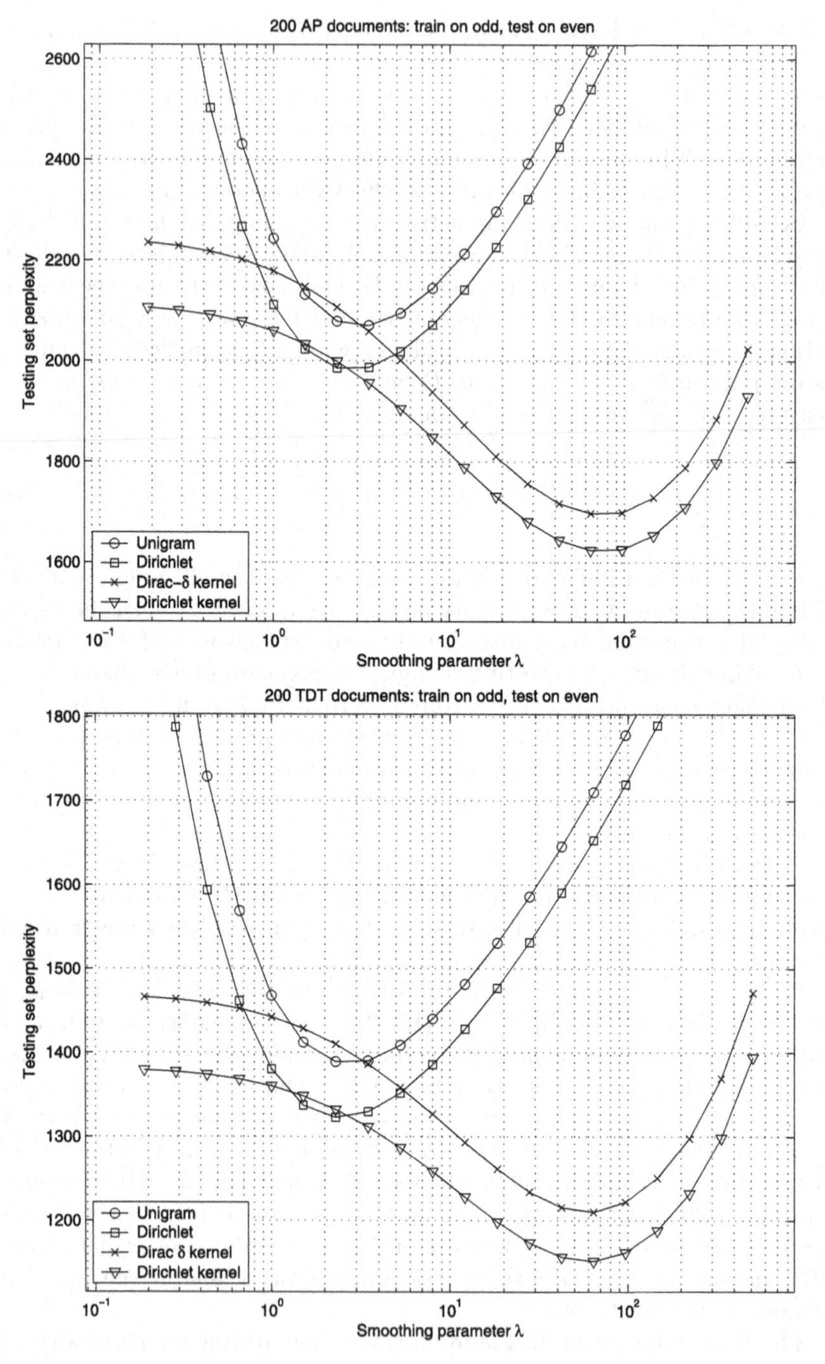

Fig. 4.9. Predictive performance of generative models on the two datasets: AP (top) and TDT (bottom). Kernel-based models outperform the baselines by 14-20%. Dirichlet kernels yield better performance than delta kernels.

on both datasets. On the AP dataset, the perplexity is reduced by 20% from 2000-2100 to 1600-1700. On the TDT dataset we observe a more modest 14% reduction from around 1400 to around 1200. The figures demonstrate that kernel-based models are well suited for capturing whetever structure is present in the word occurrences.

The second interesting observation is that models based on Dirichlet distributions are more effective than models based on delta functions. Recall that Dirichlet distributions form "bumps" over the generative simplex, and Figure 4.9 suggests that these bumps are 5-7% more effective than the "spikes" corresponding to the unigram model and the δ-kernel model. There are two equally valid explanations for this behavior. The first is that delta spikes yield a high-variance density estimate over the generative simplex. The spikes will jump around if we consider a different training subset $\mathcal{A}$. Dirichlet bumps, on the other hand, are much more stable. The peaks of these bumps may jump around, but we should not see drastic differences in the overall surface. The second explanation stems from our argument in the end of section 4.2.3. We showed that the main benefit of using a Dirichlet is that it will properly capture *word contagion*, i.e. dependencies between successive occurrences of the same word. Word repetitions are quite common in natural language, so it should not come as a complete surprise that capturing repetitions leads to better predictive performance of a model.

The third, and final observation from Figure 4.9 is that the optimal setting of the smoothing parameter λ is quite stable across the two datasets. This suggests that we should be able to pick an optimal setting on one dataset, and use it to get good performance on a different dataset.

4.6 Summary

We began this chapter by arguing for the need to have powerful generative models for exchangeable sequences of data. The need was motivated by the generative relevance hypothesis. Recall that the GRH allows us to reduce the problem of retrieval to the problem of determining whether document d and query q could be samples from the same underlying generative model $P_R(\cdot)$. Prompted by the GRH, we devoted this entire chapter to developing a powerful framework for generative models. Our development took the following steps:

1. We gave an overview of five existing generative models, starting with very basic formulas and progressing to advanced models, such as pLSI and LDA. We also argued that a limiting factor in these advanced models is the assumption of a fixed number of topics. While the assumption is very intuitive and appealing to humans, it carries no intrinsic benefit for generative modeling.
2. We demonstrated that all five models can be reduced to the following simple form:

$$P(w_1 \ldots w_n) = \int_{\mathbb{P}_{\mathcal{V}}} \prod_{i=1}^{n} u(w_i) p(du)$$

We argued that the only difference between the models is in the way they allocate density $p(du)$ to regions of the generative simplex $\mathbb{P}_{\mathcal{V}}$. In the course of our arguments we made two surprising observations: **(i)** pLSI is a special case of a simple mixture model **(ii)** LDA is equivalent to a Dirichlet model restricted to the topic sub-simplex. We also argued that looking at geometric aspects of a generative model may be more elucidating than analyzing its graphical structure.

3. We proposed a new generative model, based on a kernel-based estimate of $p(du)$. We suggested two possible kernel types for the model: delta kernels (spikes) and Dirichlet kernels (bumps). We argued that the new model has the following attractive properties: **(i)** it makes no questionable assumptions about the structure of the space, **(ii)** it is able to handle rare events and preserve outliers, and **(iii)** it is reasonably efficient and allows for easy discriminative training when necessary.
4. We demonstrated that the new generative model is effective. The model yields a 14-20% reduction in predictive perplexity over two natural baselines.

This chapter completes the formal development of our model. We now have all the pieces necessary to construct operational systems based on the generative relevance hypothesis. As a relevance model, we will use the kernel-based density allocation model with delta kernels. In some cases computational complexity will force us to fall back to the simple Dirichlet model. The following chapter will describe specific instantiations of our relevance model in various search scenarios. The chapter will also contain extensive evaluation of retrieval effectiveness for each scenario.

5

Retrieval Scenarios

In the previous chapters we tried to keep our definitions very abstract – talking about conceptual "representation" spaces, "transform" functions and unspecified "parameter" vectors. Our motivation was to keep the model as general as possible, so it would be applicable to a wide range of retrieval problems. Now is the time to bring our discussion down to earth and provide specific definitions for a number of popular retrieval scenarios. We will discuss the following retrieval scenarios:

1. **Ad-hoc retrieval:** we have a collection of English documents, and a short English query. The goal is to retrieve documents relevant to the query.
2. **Relevance feedback:** in addition to the query, the user provides us with a few examples of relevant documents. The goal is to retrieve more relevant documents.
3. **Cross-language retrieval:** we have a collection of Chinese documents and an English query. The goal is to find Chinese relevant documents.
4. **Handwriting retrieval:** we have a set of historical manuscripts, represented as bitmap images. The goal is to search the collection using text queries.
5. **Image retrieval:** we have a collection of un-labeled photographs. The goal is to identify photographs relevant to a given text query (e.g., find "tiger in the grass").
6. **Video retrieval:** we have a collection of un-annotated video footage. The goal is to find video shots containing objects of interest (e.g., "forest fire").
7. **Structured search with missing data:** we have a database with missing field values in many records. The goal is to satisfy structured queries in the face of incomplete data.
8. **Topic detection and tracking:** we have a live stream of news reports. The goal is to organize the reports according to the events discussed in them.

For each of these retrieval scenarios we will address the following three issues:

(i) **Definition.** We will discuss the nature of the retrieval scenario, the goal we are trying to achieve, and the possible variations in the scenario.
(ii) **Representation.** We will provide a detailed description of how to adapt our model to the specific scenario. The representation section will answer the following questions:
- What is the appropriate representation space $\mathcal{S}$?
- How do we represent documents and queries in that space?
- What is the probability distribution $P(\cdot)$ over $\mathcal{S}$?
- How can we estimate a relevance model from the user's query?
- How do we rank documents in response to the query?

(iii) **Experiments.** We will describe a set of experiments comparing the accuracy of the generative relevance model against state-of-the-art baseline systems.

Before we dive into the scenarios we would like to remind our reader that the representation space $\mathcal{S}$ and the transforms D and Q represent imaginary formalisms. In actuality, we always deal with documents $\mathbf{d}$ and queries $\mathbf{q}$, and never with their pre-images $\mathbf{x}$. The representation space and the transforms are introduced solely for the purpose of constructing a joint probability distribution P.

5.1 Ad-hoc Retrieval

We will start with the simplest possible scenario – ad-hoc retrieval. In this case we have a collection of documents $\mathcal{C}$, and a user's query $\mathbf{q}$. The documents in the collection have no structure and no meta-information. Aside from the query, we know nothing about the user and his information need. The query typically comes in the form of a small set of keywords, in the same language as the documents. In some cases, the user may provide a somewhat longer, syntactically well-formed request – a narrative discussing his information need. The goal of an information retrieval system is to rank all documents in the collection in such a way that relevant documents will end up at the top of the ranking. After the query is entered, no interaction is allowed to take place between the user and the system.

5.1.1 Representation

Since documents and queries share the same language, it makes sense to make that language the basis for our representation space $\mathcal{S}$. Let $\mathcal{V}$ be the vocabulary of our language and let M be an upper bound on document length. The representation space $\mathcal{S}$ will be the space of all possible sequences of length $M-1$, consisting of words from $\mathcal{V}$. The random variables $X_1 \ldots X_M$ will have the following meaning. X_1 will specify the length of the word sequence (document or query), a natural number less than M. Each subsequent variable $X_{i>1}$ will be a word from our vocabulary $\mathcal{V}$.

Document and query representation

The document generating transform D is very simple – it takes a representation $X = n, w_1...w_{M-1}$ and returns the first n words as the document:

$$D(X) = D(n, w_1...w_{M-1}) = n, w_1...w_n \tag{5.1}$$

The query transform Q is only slightly more complex. We assume that inside Q there is a mechanism that somehow decides for each word w_i whether it would make a good query word. Selected words form a subsequence $w_{i_1}...w_{i_m}$, and Q returns that sub-sequence as the query:

$$Q(X) = Q(n, w_1...w_{M-1}) = m, w_{i_1}...w_{i_m} \tag{5.2}$$

We do not have to bother with the details of that mechanism since we will always be dealing with already existing queries – for all intents and purposes that mechanism could be a deterministic human being. Since the representation space contains all possible strings over the vocabulary, we are guaranteed that D and Q can produce any observed document and any observed query.

Probability distribution over representations

The first component of all representations is a document length n, or query length m. As a first approximation, we will assume that document / query length is independent of everything and follows a uniform distribution:

$$P(X_1{=}n) = \frac{1}{M-1} \quad \text{for } n < M \tag{5.3}$$

All other variables $X_{i>1}$ take values in the same space: the vocabulary $\mathcal{V}$. A natural distribution over a vocabulary is a multinomial distribution. θ will denote the set of parameters of that multinomial, i.e. for each word v, there will be a parameter θ_v which tells us how likely we are to observe the word v. For each $i > 1$ we have:

$$P(X_i{=}v) = P_{i,\theta}(v) = \theta_v \tag{5.4}$$

Since each θ is a distribution over the $\mathcal{V}$, the space Θ of all parameter settings will be the simplex of all possible probability distributions over the vocabulary: $\Theta = \mathbb{P}_{\mathcal{V}} = \{\mathbf{x}{\in}[0,1]^{\mathcal{V}} : |x| = 1\}$. We will adopt kernel-based density allocation with delta kernels (equation 4.31). Under the above choices, the probability of observing a given query will take the following form:

$$\begin{aligned} P(Q{=}\{m, q_1...q_m\}) &= \frac{1}{M{-}1} \int_{\mathbb{P}_{\mathcal{V}}} \left\{ \prod_{i=1}^{m} u(q_i) \right\} p_{ker,\delta}(du) \\ &= \frac{1}{N(M{-}1)} \sum_{\mathbf{d}} \prod_{i=1}^{m} u_{\mathbf{d}}(q_i) \end{aligned} \tag{5.5}$$

The summation goes over all training strings **d** which define the points where we place the kernels. In the ad-hoc scenario we have no special training collection, so we will use all available documents $\mathbf{d}\in\mathcal{C}$ as training strings. The probability of observing a given document has the same form as equation (5.5), except the product goes over the words $d_1\ldots d_n$ in the document.

Relevance model

In the ad-hoc retrieval scenario the user's query $\mathbf{q}=q_1\ldots q_m$ represents the only observable sample from the relevant population. Accordingly, we estimate the relevance model $RM_\mathbf{q}$ based on the query alone. The relevance model takes the following form:

$$
\begin{aligned}
RM_\mathbf{q}(v) &= \frac{P(v, q_1\ldots q_m)}{P(q_1\ldots q_m)} \\
&= \frac{\sum_\mathbf{d} u_\mathbf{d}(v)\prod_{i=1}^m u_\mathbf{d}(q_i)}{\sum_\mathbf{d}\prod_{i=1}^m u_\mathbf{d}(q_i)}
\end{aligned} \tag{5.6}
$$

The second step of equation (5.6) comes from using equation (5.5) to represent the joint distribution of a set of words. Note that the constant factors $\frac{1}{N(M-1)}$ cancel out in the process as they are independent of all other variables. As before, summation goes over all documents **d** in the collection, and the empirical distributions $u_\mathbf{d}(\cdot)$ are estimated as follows:

$$
u_\mathbf{d}(v) = \lambda_\mathbf{d}\frac{n(v,\mathbf{d})}{|\mathbf{d}|} + (1-\lambda_\mathbf{d})BG(v) \tag{5.7}
$$

Here $n(v,\mathbf{d})$ is the count of word v in the document, $|\mathbf{d}|$ denotes the document length and $BG(v)$ is the background probability of v defined according to equation (3.15). $\lambda_\mathbf{d}$ is the smoothing parameter tuned to optimize retrieval performance.

Document ranking

We will use both the probability ratio and cross entropy divergence for ranking documents. Recall that the probability ratio, defined in section 3.6.1, was directly derived from Robertson's Probability Ranking Principle [114] and involves ranking the documents according to the following ratio:

$$
\frac{P(D{=}d_1\ldots d_n|R{=}1)}{P(D{=}d_1\ldots d_n|R{=}0)} \approx \frac{\prod_{i=1}^n RM_\mathbf{q}(d_i)}{\prod_{i=1}^n BG(d_i)} \tag{5.8}
$$

Here $RM_\mathbf{q}(\cdot)$ is the relevance model defined in equation (5.6), and $BG(\cdot)$ is the background distribution over the words, computed according to equation (3.15). As we argued before, the background distribution is going to provide a good reflection of word probabilities in the non-relevant documents (R=0).

Cross-entropy is rank-equivalent to Kullback-Leiblar divergence, which forms the basis of a hypothesis test for the event that the query and the document originated in a common population; it has been discussed extensively in section 3.6.2. We will rank the documents **d** by the cross entropy between the model of the relevant class $RM_{\mathbf{q}}$ and the empirical distribution of the words in the document $u_{\mathbf{d}}$:

$$-H(RM_{\mathbf{q}}||u_{\mathbf{d}}) = \sum_{v\in\mathcal{V}} RM_{\mathbf{q}}(v)\log u_{\mathbf{d}}(v) \stackrel{\text{rank}}{=} -KL(RM_{\mathbf{q}}||u_{\mathbf{d}}) \quad (5.9)$$

As before, $RM_{\mathbf{q}}(\cdot)$ is given by equation (5.6), and $u_{\mathbf{d}}(\cdot)$ is the empirical distribution of words in the document **d**, computed according to equation (4.29).

5.1.2 Examples of Relevance Models

Table 5.1 provides four examples of relevance models constructed from extremely short samples. We constructed the models as described in section 5.1.1, using the titles of several TDT topics as relevant samples. The titles are shown at the top of each column, and below them we show 10 words that get the highest probabilities under the relevance model. For every topic, we can see a lot of words which are highly-related to the topic, but also a few general words, perhaps unrelated to the particular topic. This is not surprising, since a relevance model is the probability distribution over all words in the relevant documents, and as such will certainly include common words. Stop-words (very frequent words, such as "of", "the", etc.) were excluded from the computation.

"monica lewinsky"	"rats in space"	"john glenn"	"unabomber"
$RM_{\mathbf{q}}(w)$ w	$RM_{\mathbf{q}}(w)$ w	$RM_{\mathbf{q}}(w)$ w	$RM_{\mathbf{q}}(w)$ w
0.041 lewinsky	0.062 rat	0.032 glenn	0.046 kaczynski
0.038 monica	0.030 space	0.030 space	0.046 unabomber
0.027 jury	0.020 shuttle	0.026 john	0.019 ted
0.026 grand	0.018 columbia	0.016 senate	0.017 judge
0.019 confidant	0.014 brain	0.015 shuttle	0.016 trial
0.016 talk	0.012 mission	0.011 seventy	0.013 say
0.015 case	0.012 two	0.011 america	0.012 theodore
0.014 president	0.011 seven	0.011 old	0.012 today
0.013 clinton	0.010 system	0.010 october	0.011 decide
0.010 starr	0.010 nervous	0.010 say	0.011 guilty

Table 5.1. Examples of mono-lingual relevance models for the TDT2 topics. Probabilities are estimated using techniques from section 5.1.1, starting with just the topic title. For each topic we show 10 words with highest estimated probabilities under the relevance model.

Table 5.1 gives us an intuition for the good performance of relevance models on the ad-hoc retrieval task. The relevance model $RM_{\mathbf{q}}$ achieves the same effect as a massive *query expansion* technique: we augment the original query $\mathbf{q}$ with thousands of statistically related words. It is important to stress that query expansion inherent in $RM_{\mathbf{q}}$ is both context-aware and domain-aware. Context awareness comes from the fact that word probabilities in the relevance model are conditioned on the query *as a whole*, not on individual query words as would be the case with synonym expansion. Context awareness provides a strong disambiguation effect: for example the query *"sandy bank"* would not be augmented with synonyms from the financial sense of the word *"bank"*. Domain-awareness happens because the relevance model is constructed over the documents in the collection we are searching, so related words will naturally reflect the peculiarities of our domain.

5.1.3 Experiments

In this section we turn our attention to evaluating the performance of the proposed retrieval model on a number of TREC collections. The experiments were originally reported in [75]. The remainder of this section is organized as follows. We start by describing the evaluation metrics and the corpora which were used in our experiments. We will then discuss the baseline systems that will be used to gauge the retrieval effectiveness of our model and provide detailed performance comparisons. Finally, at the end of this section we will compare the retrieval effectiveness of probability ratio and cross-entropy ranking criteria.

Evaluation Paradigm

The primary goal of any Information Retrieval system lies in identifying a set $\mathcal{C}_R \subset \mathcal{C}$ of documents relevant to the user's information need. This is a set-based decision task, but in practice most retrieval systems are evaluated by how well they can *rank* the documents in the collection. Let $d_1, d_2, \ldots d_N$ denote some ordering of the documents in the collection. Then, for every rank k, we can compute *recall* as the number of relevant documents that were observed in the set $\{d_1 \ldots d_k\}$, divided by the total number of relevant documents in the collection. Similarly, *precision* is defined as the number of relevant documents among $\{d_1 \ldots d_k\}$ divided by k. System performance is evaluated by comparing precision at different levels of recall, either in a form of a table (e.g. Table 5.3), or as a recall-precision graph (e.g. Figure 5.1). A common objective is to increase precision at all levels of recall. For applications that require interaction with the user, it is common to report precision at specific ranks, e.g. after 5 or 10 retrieved documents. When one desires a single number as a measure of performance, a popular choice is *average precision* defined as an arithmetic average of precision values at every rank where a relevant document occurs. Another possible choice is R-precision: precision that is achieved at rank R, where R is the number of relevant documents in the dataset. In all of these measures, precision values are usually averaged across a large set of queries with known relevant sets.

We will adopt *mean average precision* as the primary evaluation measure for all the experiments in this paper. In most cases we will also report precision at different recall levels and precision at specific ranks. When possible, we will report the results of statistical significance tests.

Name	Sources	Years	#Docs	#Terms	dl	cf	Queries	ql
AP	Associated Press	89-90	242,918	315,539	273	210	51-150	4.32
FT	Financial Times	91-94	210,158	443,395	237	112	251-400	2.95
LA	Los Angeles Times	89-90	131,896	326,609	290	117	301-400	2.52
WSJ	Wall Street Journal	87-92	173,252	185,903	265	247	1-200	4.86
TDT2	AP, NYT, CNN, ABC, PRI, VOA	1998	62,596	122,189	167	85	TDT	3.02

Table 5.2. Information for the corpora used in ad-hoc retrieval experiments. *dl* denotes average document length, *cf* stands for average collection frequency of a word, and *ql* represents average number of words per query.

Datasets and processing

We use five different datasets in our evaluation of adhoc retrieval effectiveness. Table 5.2 provides detailed information for each dataset. All five datasets contain news releases; the majority of them are print media, although the TDT2 corpus contains a significant broadcast component. The datasets vary in size, time frame, and word statistics. All datasets except TDT2 are homogeneous, i.e. they contain documents from a single source. For each dataset there is an associated set of topics, along with human *relevance judgments*. For the TDT2 dataset the judgments are exhaustive, meaning that every document has been manually labeled as either relevant or non-relevant for every topic. The other four datasets contain *pooled* judgments, i.e. only top-ranked documents from a set of retrieval systems were judged with respect to each topic by annotators at NIST. TREC topics come in the form of queries, containing title, description and narrative portions. We used only the titles, resulting in queries which are 3-4 words in length. TDT2 topics are defined by a set of examples and do not have associated queries. However, short titles were assigned to them by annotators, and we used these titles as queries.

Prior to any experiments, each dataset was processed as follows. Both documents and queries were tokenized on whitespace and punctuation characters. Tokens with fewer than two characters were discarded. Tokens were then lower-cased and reduced to their root form by applying the Krovetz stemmer used in the InQuery engine [3]. The stemmer combines morphological rules with a large dictionary of special cases and exceptions. After stemming, 418 stop-words from the standard InQuery [3] stop-list were removed. All of the remaining tokens were used for indexing, and no other form of processing was used on either the queries or the documents.

Baseline systems

We will compare retrieval performance of generative relevance models (**RM**) against four established baselines:

tf.idf Our first baseline represents one of the most widely-used and successful approaches to adhoc retrieval. We use the InQuery [3] modification of the popular Okapi *BM25* [119] weighting scheme. Given a query $Q = q_1 \ldots q_k$, the documents D are ranked by the following formula:

$$S(Q,D) = \sum_{q \in Q} \frac{n(q,D)}{n(q,D) + 0.5 + 1.5\frac{dl}{avg.dl}} \frac{\log(0.5 + N/df(q))}{\log(1.0 + \log N)} \tag{5.10}$$

Here $n(q,D)$ is the number of times query word q occurs in document D, dl is the length of document D, $df(q)$ denotes the number of documents containing word q, and N stands for the total number of documents in the collection. Equation (5.10) represents a heuristic extension of the classical probabilistic models of IR, discussed in section 2.3.2. The formula is remarkable for its consistently good performance in yearly TREC evaluations.

LCA From an IR standpoint, our retrieval model contains a massive *query expansion* component (see section 5.1.2). To provide a fair comparison, we describe performance of a state-of-the-art heuristic query expansion technique: Local Context Analysis (LCA) [151]. LCA is a technique for adding highly-related words to the query, in the hope of handling synonymy and reducing ambiguity. Given a query $Q = q_1 \ldots q_k$, and a set of retrieved documents R, LCA ranks all words w in the vocabulary by the following formula:

$$Bel(w;Q) = \prod_{q \in Q} \left(0.1 + \frac{1/\log n}{1/idf_w} \log \sum_{D \in R} D_q D_w \right)^{idf_q} \tag{5.11}$$

Here n is the size of the retrieved set, idf_w is the inverse document frequency [117] of the word w; D_q and D_w represent frequencies of words w and q in document D. To perform query expansion, we add m highest-ranking words to the original query $q_1 \ldots q_k$, and perform retrieval using **tf.idf** method described above. n and m represent parameters that can be tuned to optimize performance. Based on preliminary experiments, we set $n = 10$ and $m = 20$.

LM Our third baseline represents the language-modeling framework, pioneered by Ponte and Croft [106] and further developed by a number of other researchers [127, 56, 90]. The framework was discussed in detail in section 2.3.3. Recall that in the language-modeling framework we view the query $Q = q_1 \ldots q_k$ as an i.i.d. random sample from some unknown model representing a perfectly relevant document. Documents D are ranked by the probability that $q_1 \ldots q_k$ would be observed during random sampling from the model of D:

$$P(Q|D) = \prod_{q \in Q} \left(\lambda_D \frac{n(q, D)}{dl} + (1 - \lambda_D) P(q) \right) \tag{5.12}$$

Here $n(q, D)$ is the number of times q occurs in D, dl is the length of D, and $P(q)$ is the background probability of the query word q, computed over the entire corpus. The smoothing parameter λ_D was set to the Dirichlet [157] estimate $\lambda_D = \frac{dl}{dl+\mu}$. Based on previous experiments, μ was set to 1000 for all TREC corpora and to 100 for the TDT2 corpus.

Recall from section 3.6.3 that this method of ranking is equivalent to using relative entropy where the relevance model was replaced with the empirical distribution of words in Q.

LM+X The language-modeling framework does not contain an inherent query expansion component. In his thesis, Ponte [105] developed a heuristic query expansion approach that demonstrated respectable performance when combined with the ranking formula described above. Given the set of documents R, retrieved in response to the original query, we rank all the words w by the following formula:

$$Bel(w; R) = \sum_{D \in R} log \left(\frac{P(w|D)}{P(w)} \right) \tag{5.13}$$

Here $P(w|D)$ is the smoothed relative frequency of word w in D, and $P(w)$ is the background probability of w. m highest-ranking words are added to the original query $q_1 \ldots q_k$, and retrieval was performed according to the **LM** method described above. Based on prior experiments, we added m=5 words from the n=5 top-ranked documents.

Comparing the generative relevance model to baselines

Table 5.3 presents the performance of five systems on the Wall-Street Journal (WSJ) dataset with TREC title queries 1 - 200. We observe that in general, language-modeling approaches (**LM** and **LM+X**) are slightly superior to their heuristic counterparts. Query expansion techniques lead to significant improvements at high recall. Relevance models (**RM**) noticeably outperform all four baselines at all levels of recall, and also in terms of average precision and R-precision (precision at the number of relevant documents). Improvements over **tf.idf** are all statistically significant at the 95% confidence level according to the Wilcoxon signed-rank test [147].

WSJ: TREC queries 1-200 (title)

	tf.idf	LCA	%chg	LM	%chg	LM+X	%chg	RM	%chg
Rel	20982	20982		20982		20982		20982	
Rret	11798	12384	4.97*	12025	1.92*	12766	8.20*	13845	17.35*
0.00	0.634	0.666	4.9	0.683	7.6*	0.653	3.0	0.686	8.1*
0.10	0.457	0.465	1.8	0.474	3.8	0.481	5.4	0.533	16.8*
0.20	0.383	0.380	-0.8	0.395	3.0	0.403	5.1	0.463	20.7*
0.30	0.334	0.333	-0.3	0.340	1.8	0.352	5.3	0.403	20.6*
0.40	0.287	0.283	-1.4	0.288	0.2	0.307	6.7	0.350	21.9*
0.50	0.240	0.240	0.1	0.246	2.7	0.270	12.6*	0.304	26.6*
0.60	0.191	0.195	2.3	0.203	6.5	0.226	18.5*	0.254	33.0*
0.70	0.138	0.153	11.1*	0.158	14.7*	0.178	29.0*	0.196	42.2*
0.80	0.088	0.108	22.8*	0.110	24.9*	0.133	50.7*	0.146	66.1*
0.90	0.049	0.061	25.3*	0.074	51.8*	0.080	64.2*	0.085	73.3*
1.00	0.011	0.013	25.4	0.017	55.6*	0.022	104.3*	0.022	99.7*
Avg	0.238	0.244	2.89	0.253	6.61*	0.265	11.46*	0.301	26.51*
5	0.4360	0.4470	2.5	0.4480	2.8	0.4620	6.0*	0.5170	18.6*
10	0.4060	0.4185	3.1	0.4140	2.0	0.4220	3.9	0.4675	15.1*
15	0.3830	0.3827	-0.1	0.3900	1.8	0.3963	3.5	0.4463	16.5*
20	0.3662	0.3580	-2.3	0.3737	2.0	0.3805	3.9	0.4258	16.2*
30	0.3390	0.3277	-3.3	0.3498	3.2*	0.3483	2.8	0.3933	16.0*
100	0.2381	0.2318	-2.7	0.2432	2.1*	0.2513	5.5	0.2744	15.3*
200	0.1725	0.1713	-0.7	0.1760	2.1*	0.1822	5.6*	0.1996	15.7*
500	0.0978	0.1018	4.1*	0.1009	3.1*	0.1041	6.4*	0.1138	16.3*
1000	0.0590	0.0619	5.0*	0.0601	1.9*	0.0638	8.2*	0.0692	17.4*
RPr	0.2759	0.2753	-0.23	0.2835	2.73*	0.2871	4.03	0.3162	14.58*

Table 5.3. Comparison of Relevance Models (RM) to the baseline systems: (tf.idf) Okapi / InQuery weighted sum, (LCA) tf.idf with Local Context Analysis, (LM) language model with Dirichlet smoothing, (LM+X) language model with Ponte expansion. Relevance Models noticeably outperform all baseline systems. Stars indicate statistically significant differences in performance with a 95% confidence according to the Wilcoxon test. Significance tests are performed against the tf.idf baseline.

In Tables 5.4 and 5.5 we show the performance on the remaining four datasets: AP, FT, LA and TDT2. We compare relevance models (**RM**) with two of the four baselines: **tf.idf** and **LM**. As before, we notice that language modeling approach (**LM**) is somewhat better than the heuristic **tf.idf** ranking. However, the improvements are not always consistent, and rarely significant. Relevance models demonstrate consistent improvements over both baselines on all four datasets. Compared to **tf.idf**, overall recall of relevance models is higher by 12% - 25%, and average precision is up by 15% - 25%. Improvements are statistically significant with 95% confidence according to the Wilcoxon test.

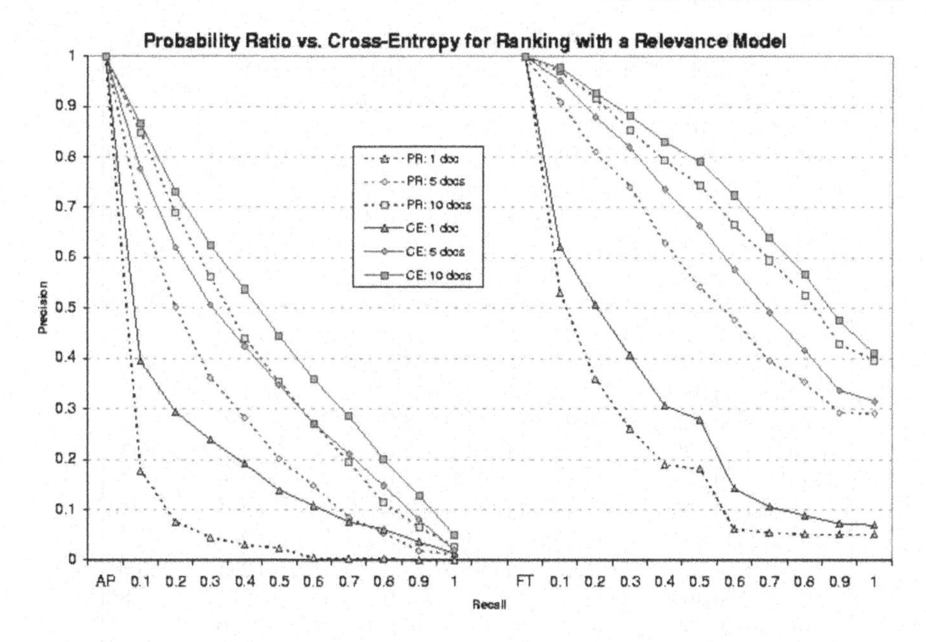

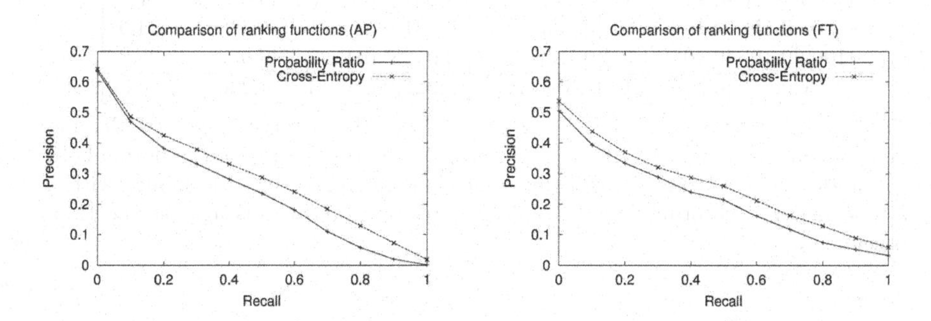

Fig. 5.1. Comparing the effectiveness of two ranking functions: probability ratio (PR) and cross-entropy (CE). We show results for two datasets: (AP) on the left side and (FT) on the right. Using cross-entropy leads to noticeably higher precision at all levels of recall. The difference is consistent for relevance models of different quality: 1, 5 or 10 training documents (top) as well as relevance models estimated from the words in the query.

	AP: TREC queries 51-150 (title)					FT: TREC queries 251-400 (title)				
	tf.idf	LM	%chg	RM	%chg	tf.idf	LM	%chg	RM	%chg
Rel	21809	21809		21809		4816	4816		4816	
Ret	10115	10137	0.2	12525	23.8*	2541	2593	2.0*	3197	25.8*
0.0	0.644	0.643	-0.1	0.632	-1.8	0.531	0.561	5.5	0.535	0.8
0.1	0.442	0.436	-1.4	0.484	9.4*	0.415	0.421	1.4	0.430	3.7
0.2	0.359	0.349	-2.8	0.425	18.4*	0.353	0.355	0.4	0.368	4.4
0.3	0.308	0.299	-3.0*	0.379	23.1*	0.291	0.303	4.3	0.316	8.8*
0.4	0.255	0.246	-3.5	0.333	30.8*	0.249	0.258	3.6	0.282	13.3*
0.5	0.212	0.209	-1.4	0.289	35.9*	0.213	0.230	7.8	0.256	20.0*
0.6	0.176	0.170	-3.6*	0.246	39.8*	0.158	0.187	17.9*	0.210	32.6*
0.7	0.128	0.130	1.3	0.184	43.9*	0.108	0.137	26.4*	0.160	47.3*
0.8	0.084	0.086	2.8	0.128	52.0*	0.078	0.102	30.6*	0.128	63.3*
0.9	0.042	0.048	14.9	0.071	70.8*	0.058	0.078	32.9*	0.089	53.0*
1.0	0.016	0.022	38.8*	0.018	10.5	0.042	0.066	56.5*	0.059	40.1*
Avg	0.222	0.219	-1.4	0.277	25.0*	0.211	0.230	8.8*	0.246	16.7*
5	0.430	0.457	6.1	0.481	11.7*	0.306	0.332	8.5	0.331	8.0
10	0.420	0.434	3.4	0.470	11.8*	0.263	0.276	4.9	0.283	7.7
15	0.410	0.417	1.6	0.457	11.3*	0.224	0.251	12.2*	0.244	9.0*
20	0.396	0.409	3.4	0.448	13.3*	0.201	0.225	12.0*	0.218	8.8*
30	0.380	0.390	2.7	0.429	12.9*	0.170	0.187	10.2*	0.190	11.7*
100	0.302	0.305	0.8	0.354	17.0*	0.095	0.099	4.4*	0.111	17.0*
200	0.240	0.242	1.2	0.294	22.6*	0.061	0.064	5.2*	0.074	21.7*
500	0.156	0.157	0.9	0.193	23.7*	0.031	0.032	1.6*	0.039	24.9*
1000	0.102	0.102	0.2	0.127	23.8*	0.018	0.019	2.0*	0.023	25.8*
RPr	0.272	0.268	-1.6	0.315	15.7*	0.227	0.236	3.8*	0.236	4.0

Table 5.4. Comparison of Relevance Models (RM) to the InQuery (tf.idf) and language-modeling (LM) systems. Relevance Model significantly outperforms both baselines. Stars indicate statistically significant differences in performance with a 95% confidence according to the Wilcoxon test. Significance is against the tf.idf baseline.

Comparing different document ranking criteria

In section 3.6 we discussed two ways of ranking documents in response to the query: the probability ratio, based on Robertson's probability ranking principle, and Kullback-Leiblar divergence, which forms the bases for a χ^2 hypothesis test. In section 3.6.3 we argued that KL-divergence may be analytically superior to the probability ratio, even if the relevance model can be estimated perfectly. In this section we provide empirical support for our arguments.

In our comparison, we want to factor out the effects of what method we use for estimating the relevance model. So we will compare both the *query-induced* relevance models and the models estimated from genuine *relevant examples*. We start by estimating the model RM_r from a set of 1, 5 or 10 training documents. In each case we use smoothed maximum-likelihood mod-

	LA: TREC queries 301-400 (title)					TDT2: TDT topics 1-100 (title)				
	tf.idf	LM	%chg	RM	%chg	tf.idf	LM	%chg	RM	%chg
Rel	2350	2350		2350		7994	7994		7994	
Ret	1581	1626	2.9	1838	16.3*	5770	5399	-6.4	6472	12.2*
0.0	0.586	0.619	5.7	0.566	-3.5	0.846	0.843	-0.4	0.854	0.9
0.1	0.450	0.486	8.2*	0.474	5.5	0.794	0.797	0.4	0.831	4.5*
0.2	0.356	0.362	1.9	0.384	8.1	0.755	0.748	-0.9	0.806	6.8*
0.3	0.295	0.316	7.0*	0.332	12.5*	0.711	0.705	-0.9	0.785	10.3*
0.4	0.247	0.273	10.4	0.286	15.6*	0.663	0.669	0.8	0.766	15.5*
0.5	0.217	0.238	9.8	0.264	21.7*	0.614	0.616	0.3	0.742	20.9*
0.6	0.164	0.197	19.7	0.206	25.5*	0.565	0.563	-0.5	0.704	24.6*
0.7	0.129	0.159	23.0*	0.174	34.9*	0.528	0.517	-2.1	0.675	27.8*
0.8	0.100	0.123	23.1	0.138	37.5*	0.485	0.477	-1.5*	0.648	33.6*
0.9	0.050	0.074	47.0	0.085	69.4*	0.397	0.394	-0.7*	0.587	48.0*
1.0	0.042	0.059	41.5*	0.056	32.6*	0.297	0.307	-3.5	0.477	60.4*
Avg	0.223	0.247	10.6*	0.258	15.6*	0.596	0.592	-0.6	0.709	18.9*
5	0.331	0.337	1.9	0.357	8.0	0.590	0.598	1.4	0.671	13.8*
10	0.274	0.282	3.0	0.299	9.3	0.538	0.543	0.8	0.607	12.8*
15	0.237	0.250	5.7	0.262	10.6	0.485	0.502	3.6*	0.556	14.6*
20	0.213	0.220	3.4	0.228	7.2	0.453	0.457	0.9	0.508	12.2*
30	0.179	0.186	4.0	0.195	8.7	0.402	0.398	-0.9	0.453	12.8*
100	0.092	0.094	2.3	0.102	11.8	0.247	0.243	-1.1	0.288	17.0*
200	0.054	0.057	5.2	0.066	20.2*	0.169	0.165	-2.4*	0.193	14.2*
500	0.028	0.029	5.9	0.033	18.9*	0.095	0.092	-3.8	0.109	14.4*
1000	0.016	0.017	2.8	0.019	16.3*	0.060	0.056	-6.4	0.067	12.2*
RPr	0.235	0.257	9.1*	0.251	6.5	0.564	0.567	0.51	0.672	19.3*

Table 5.5. Comparison of Relevance Models (RM) to the InQuery (tf.idf) and language-modeling (LM) systems. Relevance Model significantly outperforms both baselines. Stars indicate statistically significant differences in performance with a 95% confidence according to the Wilcoxon test. Significance is against the tf.idf baseline.

els as described in section. Smoothing parameters were tuned individually to maximize performance. The quality of the resulting rankings are shown in the top portion of Figure 5.1. Solid lines reflect the performance of rankings based on cross-entropy (which is rank-equivalent to Kullback-Leiblar divergence), dashed lines correspond to the probability ratio. We observe two consistent effects. First, as expected, more training documents translates to better performance. Relevance models estimated from 5 documents exhibit higher precision than models constructed from 1 relevant document, using 10 documents leads to an even better performance. Second, rankings based on cross-entropy (CE) noticeably outperform the probability ratio (PR). The precision of cross-entropy rankings is higher at all levels of recall, regardless of the quality of the relevance model. The results are consistent across two datasets: Associated Press and Financial Times. Note that relevant documents used to estimate the models were not excluded from the ranking, so precision numbers are somewhat higher than would be expected.

We observe similar results if we compare the two ranking methods with relevance models estimated without any examples of relevant documents. In this case we start with a short query and use equation (5.6), as discussed in section 5.1.1. Once the model is computed we rank the documents **d** using the cross-entropy (CE) and the probability ratio (PR). Results are presented in the lower half of Figure 5.1. As before, we observe that ranking by cross-entropy results in noticeably higher precision at all levels of recall. The difference is consistent across the two datasets involved.

A summary of the empirical comparison of the two ranking methods is provided in Table 5.6. We report mean average precision on five different datasets. In every case cross-entropy leads to substantially superior performance. The improvement is particularly dramatic on the Wall Street Journal corpus with TREC queries 1-200. Improvements are statistically significant.

Corpus	Probability ratio	Cross-entropy	% change
AP	0.228	0.272	+19%
FT	0.212	0.244	+15%
LA	0.226	0.254	+12%
WSJ	0.195	0.300	+54%
TDT2	0.621	0.712	+15%

Table 5.6. Contrasting performance of probability-ratio and cross-entropy ranking. Cross-entropy results in substantially better performance on all five datasets involved.

5.2 Relevance Feedback

Relevance feedback represents a natural extension of the ad-hoc searching scenario. In addition to the query **q**, the user is going to provide us with a small set of documents **r** he considers relevant to his information need. Our goal is to use the relevant documents to construct a better estimate of the relevance model $RM_{\mathbf{r}}(\cdot)$, which will hopefully lead to more effective document ranking. Relevance feedback algorithms have a number of practical applications. Intuitions gained from relevance feedback experiments can serve a good purpose in the areas of text classification, information routing and filtering, and Topic Detection and Tracking.

5.2.1 Representation

As far as representation is concerned, relevance feedback is no different from ad-hoc retrieval. As before, the representation X will consist of string length, followed by $M-1$ words from the vocabulary. Document and query generating transforms will be defined exactly as in the ad-hoc case (equations 5.1 and 5.2). We will use the same modeling choices, making the string length uniformly distributed over $1\ldots M-1$, and taking the parameter space Θ to be the space of all unigram distributions over the vocabulary. We will also use the same ranking criterion: cross-entropy, as given by equation (5.9). The only difference from the ad-hoc case will come in the way we estimate the relevance model $RM_{\mathbf{q}}(\cdot)$.

Previously, the relevance model was estimated from a single example $\mathbf{q}$. Now, instead of a single random sample we have a set of k random samples, denote them $\mathbf{r}^1\ldots\mathbf{r}^k$. We do not assume that these samples are homogeneous and talk about the same topic. In fact, each sample could potentially discuss a different aspect of the underlying information need. We will in turn treat each $\mathbf{r}^i$ as a sample and construct the corresponding relevance model $RM_{\mathbf{r}^i}$. During ranking, the final relevance model will be represented by the *mean* relevance model $RM_{\mathbf{r}^1\ldots\mathbf{r}^k}$, defined as follows:

$$\begin{aligned} RM_{\mathbf{r}^1\ldots\mathbf{r}^k}(v) &= \frac{1}{k}\sum_{i=1}^{k} RM_{\mathbf{r}^i}(v) \\ &= \frac{1}{k}\sum_{i=1}^{k} \frac{\sum_{\mathbf{d}} u_{\mathbf{d}}(v)\prod_{j=1}^{m_i} u_{\mathbf{d}}(\mathbf{r}^i_j)}{\sum_{\mathbf{d}}\prod_{j=1}^{m_i} u_{\mathbf{d}}(\mathbf{r}^i_j)} \end{aligned} \tag{5.14}$$

Here $\mathbf{r}^i_j$ denotes the j'th word in the i'th relevant example $\mathbf{r}^i$, and m_i is the total number of words in $\mathbf{r}^i$. Once we have the mean relevance model from equation (5.14), we simply plug it into our favorite ranking criterion, be it probability ratio (equation 5.8) or cross-entropy (equation 5.9).

We would like to point out that equation (5.14) exhibits rather peculiar behavior when the relevant examples $\mathbf{r}^1\ldots\mathbf{r}^k$ are included in the summation that goes over $\mathbf{d}$. Recall that $\sum_{\mathbf{d}}$ goes over all documents in our collection $\mathcal{C}$. If relevant examples are included in that summation, equation (5.14) becomes equivalent to the following very simple form:

$$\begin{aligned} RM_{\mathbf{r}^1\ldots\mathbf{r}^k}(v) &= \frac{1}{k}\sum_{i=1}^{k}\sum_{\mathbf{d}} u_{\mathbf{d}}(v)\left(\frac{\prod_j u_{\mathbf{d}}(\mathbf{r}^i_j)}{\sum_{\mathbf{d}}\prod_j u_{\mathbf{d}}(\mathbf{r}^i_j)}\right) \\ &\approx \frac{1}{k}\sum_{i=1}^{k}\sum_{\mathbf{d}} u_{\mathbf{d}}(v)\cdot\delta(\mathbf{d},\mathbf{r}^i) \\ &= \frac{1}{k}\sum_{i=1}^{k} u_{\mathbf{r}^i}(v) \end{aligned} \tag{5.15}$$

The reason for this curious behavior is the fact that documents are *very long* strings of text. A typical document $\mathbf{r}^i$ in any TREC collection will easily contain several thousand words (including repetitions). This means that the product $\prod_j u_{\mathbf{d}}(\mathbf{r}^i_j)$ in equation (5.15) will involve multiplying thousands of rather small numbers, leading to infinitesimally small values. Of course, the denominator will also become infinitesimally small. The effect, aside from computational nuisance, is that the ratio inside the parenthesis becomes extremely peaked: it is nearly 1 when $\mathbf{d}$ is most similar to $\mathbf{r}^i$, and nearly zero for all other points $\mathbf{d}$. In other words, it starts to act as a delta function $\delta(\mathbf{d}, \mathbf{r}^i)$. The result is that equation (5.15) simply picks out the relevant documents from the collection $\mathcal{C}$ and averages their empirical distributions. Note that the same does not happen in the ad-hoc scenario because: (i) queries are orders of magnitude shorter than relevant documents, and (ii) queries are rarely included as a kernel point $\mathbf{d}$ in the summation $\sum_{\mathbf{d}}$.

There are both positive and negative consequences stemming from equation (5.15). The upside is that we get a very simple formula for computing the expected relevant parameter vector. And for large values of k the formula is quite effective. The downside is that we will not see the benefits from the query-expansion aspect of relevance models discussed in section 5.1.2: any word that did not occur in the k relevant examples will be assigned a background probability under equation (5.15). This observation will become particularly important when we dive into the Topic Detection and Tracking scenario.

5.2.2 Experiments

In this section we will briefly report experimental results for very simple applications of user feedback in the framework of relevance models. In the experiments that follow we assumed a very simplistic feedback setting. We assume that along with the query we get a set of 5 relevant documents, chosen chronologically from the beginning of the collection. We compare the following retrieval algorithms:

System description

1. **Language Model Baseline.** Use just the original query $Q = q_1 \ldots q_m$, perform retrieval using the standard language-modeling approach with smoothing based on the Dirichlet prior [158], and the smoothing parameter set to $\mu = 1000$.
2. **Language Model Feedback.** Use Ponte's formulation of relevance feedback for language models [105]. Apply equation (5.13) to the provided set of relevant documents. Select 5 words $w_1 \ldots w_5$ with the highest average log-likelihood ratio. Add these words to the original query to form the expanded query $Q^{exp} = q_1 \ldots q_m, w_1 \ldots w_5$. Use the expanded query to perform standard language model retrieval, using Jelinek-Mercer smoothing [158] with the smoothing parameter set to $\lambda = 0.9$.
3. **Relevance Model Feedback.** Estimate a relevance model as described in section (5.2.1). Rank the documents using the relative entropy between

the relevance model and the empirical document model (equation 5.9). We used Jelinek-Mercer smoothing and the parameter λ was set to 0.1 for the document language model. Note that the original query was completely ignored.

Experimental results

Figure 5.2 shows the standard 11-point recall-precision curves for the above three algorithms. Experiments were performed on three datasets: Associated Press (AP), Financial Times (FT) and Los-Angeles Times (LA), refer to Table 5.2 for details. Five documents used for feedback were removed from evaluation in all cases.

We observe that on all three datasets both of the feedback algorithms noticeably outperform the strong language-modeling baseline. Precision is better at all levels of recall, and average precision is improved by 20-25%, which is considered standard for relevance feedback. Differences are statistically significant. If we compare the two feedback algorithms to each other, we see that their performance is almost identical. The differences are very small and not statistically significant. Note that the algorithm using relevance models completely ignored the original query, whereas Ponte's feedback used it as a basis for the expanded query. We believe the performance of the relevance modeling algorithm can be further improved by retaining the query and making use of the techniques we discussed in the ad-hoc scenario.

5.3 Cross-Language Retrieval

In a cross-language scenario, our documents are written in language A. The native language of the user is B, and consequently the queries are issued in language B. For simplicity, let's assume A stands for Arabic and B stands for Bulgarian. Our goal is to start with a Bulgarian query and return relevant Arabic documents. This may seem like an odd scenario, but it does have a number of practical applications, and has attracted a significant amount of research over the years. To give some justification for this scenario, we envision that the user has access to a human translator or to a machine translation engine, which can translate a limited amount of text from Arabic into Bulgarian. However, the Arabic database may not be available for translation as a whole, or the cost (monetary or computational) of translating every document may be too high. A natural solution is to perform cross-language retrieval and then translate only the retrieved documents.

5.3.1 Representation

As in the ad-hoc case, we assume that neither documents nor queries have any structure to them: documents are natural-language strings over the Arabic vocabulary $\mathcal{V}_A$, queries are short strings over the Bulgarian vocabulary

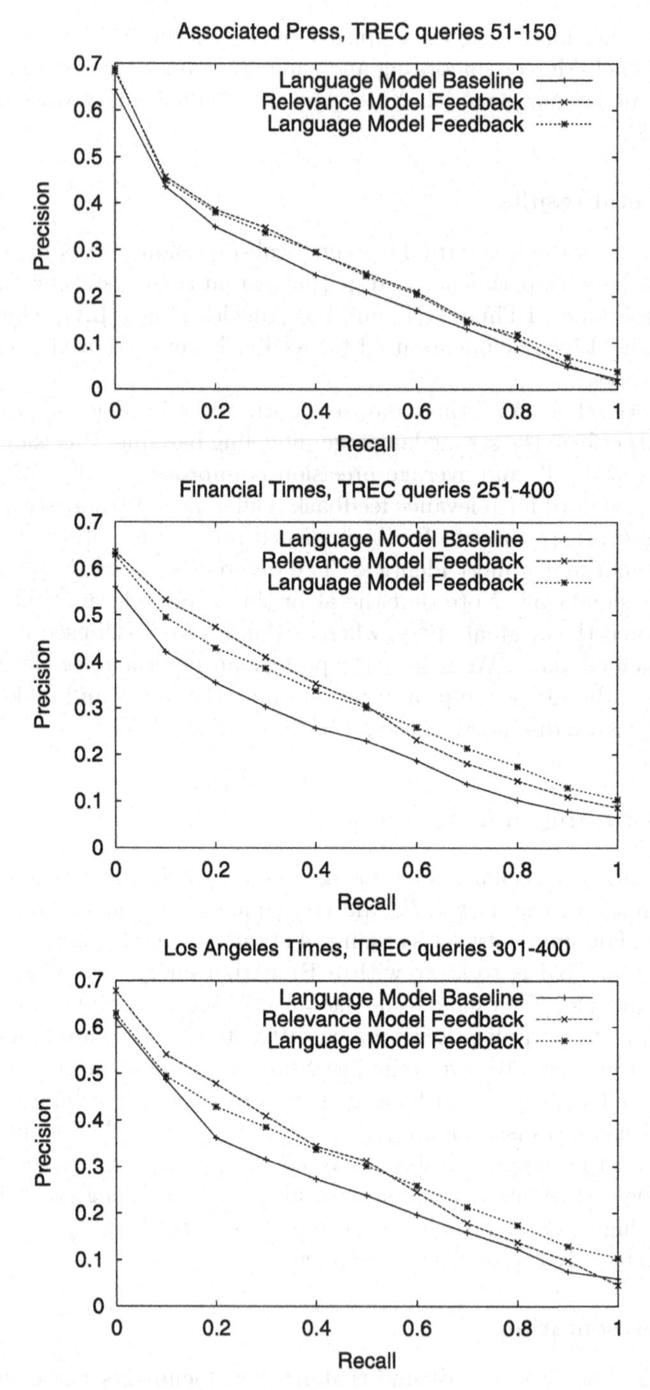

Fig. 5.2. Relevance Feedback: relevance model performs as well as the feedback mechanism proposed by Ponte. Both feedback methods significantly outperform a strong baseline.

$\mathcal{V}_B$. We assume that in its latent, unobserved form, each Arabic document comes with a ready translation into Bulgarian. So, in effect, each document is represented as two strings: one in Arabic, another in Bulgarian. The same is true for queries: each query originates as a long narrative in Bulgarian, along with its translation to Arabic. Let M_A be the upper bound on the length of Arabic strings, let M_B be a similar upper bound for Bulgarian. A latent representation of documents and queries will take the following common form: $X = n_A, a_1 \ldots a_{M_A}, n_B, b_1 \ldots b_{M_B}$. Here n_A and n_B represent the target length of description in Arabic and Bulgarian languages respectively. a_i represent all the words in the Arabic description, and b_j stand for the words from the corresponding description in Bulgarian. The entire representation space is: $\mathcal{S} = \{1 \ldots M_A\} \times \mathcal{V}_A^{M_A} \times \{1 \ldots M_B\} \times \mathcal{V}_B^{M_B}$.

Document and query representation

The document and query generating transforms will then be defined as follows:

$$D(X) = D(n_A, a_1 \ldots a_{M_A}, n_B, b_1 \ldots b_{M_B}) = n_A, a_1 \ldots a_{n_A} \tag{5.16}$$

$$Q(X) = Q(n_A, a_1 \ldots a_{M_A}, n_B, b_1 \ldots b_{M_B}) = m, b_1 \ldots b_m \tag{5.17}$$

The document transform D simply returns the Arabic portion of the latent representation, truncated to the appropriate length. The query transform Q discards the Arabic portion entirely and selects m keywords from the Bulgarian portion. Keyword selection is performed by the same hypothetical mechanism that was employed in the ad-hoc retrieval scenario (section 5.1.1).

Joint probability distribution

We assume that length of the Arabic description is uniformly distributed over $\{1 \ldots M_A\}$, and similarly length of the Bulgarian description is uniform over $\{1 \ldots M_B\}$. As in the ad-hoc scenario, we will use multinomial distributions as a natural model for words in the Arabic and Bulgarian parts of the representation. What makes the cross-language case different is the fact that we have two different vocabularies: $\mathcal{V}_A$ and $\mathcal{V}_B$. These vocabularies cannot be collapsed into one, since we want to keep the Bulgarian and Arabic strings separate. As a result, we will have to define two separate multinomial distributions for each point in the parameter space. θ will be a concatenation of two vectors: $\theta = \{\alpha, \beta\}$. The first vector will have dimensions $\alpha(a)$, defining the probabilities for every Arabic word a. The second vector will have dimensions $\beta(b)$, one for each Bulgarian word b. The space of all possible parameter settings will be the cross-product: $\Theta = \mathbb{P}_{\mathcal{V}_A} \times \mathbb{P}_{\mathcal{V}_B}$, where $\mathbb{P}_{\mathcal{V}_A} = \{\mathbf{x} \in [0,1]^{\mathcal{V}_A} : \sum_v x_v = 1\}$ is the simplex of all possible distributions over the Arabic words, and $\mathbb{P}_{\mathcal{V}_B}$ is the corresponding simplex for the Bulgarian vocabulary. As in the ad-hoc case, we will use a kernel-based density p_{ker} over the parameter space Θ, and will again opt for delta kernels (equation 4.30). Under the above definitions we will get the following expression for the probability of observing some information item:

$$\begin{aligned} &P(n_A, a_{1\dots M_A}, n_B, b_{1\dots M_B}) \\ &= \frac{1}{M_A}\frac{1}{M_B}\int_{\mathbb{P}_{\mathcal{V}_A}\times\mathbb{P}_{\mathcal{V}_B}} \left(\prod_{i=1}^{n_A}\alpha(a_i)\right)\left(\prod_{i=1}^{n_B}\beta(b_i)\right) p_{ker,\delta}(d\alpha\times d\beta) \\ &= \frac{1}{M_A}\frac{1}{M_B}\frac{1}{N}\sum_{\mathbf{w}}\left(\prod_{i=1}^{n_A}\alpha_{\mathbf{w}}(a_i)\right)\left(\prod_{i=1}^{n_B}\beta_{\mathbf{w}}(b_i)\right) \end{aligned} \tag{5.18}$$

The summation goes over every training example $\mathbf{w}$. N is the total number of training strings. The construction of the training set will be discussed below. To get the probability of observing a document D=$\mathbf{d}$ we simply restrict equation (5.18) to the Arabic part of the space by removing M_B and the product over the Bulgarian words b_i. To get the probability of a Bulgarian query, we restrict the product to just the query terms. Both of these steps are justified by our arguments in section 3.4.2.

Parameter estimation with a parallel corpus

The core part of the generative model given by equation (5.18) is a set of distributions $\alpha_{\mathbf{w}}(\cdot)$ and $\beta_{\mathbf{w}}(\cdot)$. These are supposed to be empirical distributions corresponding to some collection of training examples. $\alpha_{\mathbf{w}}(\cdot)$ is a distribution over the Arabic vocabulary, $\beta_{\mathbf{w}}(\cdot)$ is the corresponding distribution over Bulgarian words. Note that these distributions are tied to each other: they reflect word usage in the same training example $\mathbf{w}$. Such training examples are certainly consistent with our representation space $\mathcal{S}$: recall that we assumed that the latent form of each document contains both the Arabic and the Bulgarian narrative. But in reality the documents in our collection $\mathcal{C}$ are written in Arabic, the Bulgarian translation simply will not be present. Consequently, we cannot use our collection $\mathcal{C}$ as a set of training examples. What we need is a collection that does include both Arabic and Bulgarian renditions of the same conceptual narrative. Such collections are called *parallel* or *comparable* corpora, and are the staple of research in statistical machine translation [17]. There is a slight difference between parallel and comparable corpora. The former would contain an *exact* translation of each Arabic sentence into its Bulgarian equivalent. The latter does not require that sentences be exact translations, as long as groups of sentences (documents) in Arabic discuss roughly the same topical content as groups of sentences in Bulgarian. Comparable corpora are somewhat easier to obtain, but they cannot be used for training machine translation systems. Comparable corpora are sufficient for our purposes, since we are interested in topical correlations across the languages, and not in surface translation.

For the purposes of computing equation (5.18), we assume that we have access to a comparable corpus $\mathcal{C}_{train}$. This corpus consists of paired strings

$\mathbf{w} = \{\mathbf{a}, \mathbf{b}\}$, where $\mathbf{a}$ is some Arabic document and $\mathbf{b}$ is a corresponding Bulgarian rendition. Empirical distributions $\alpha_{\mathbf{w}}$ and $\beta_{\mathbf{w}}$ corresponding to a specific pair $\mathbf{w}$ would be defined as:

$$\alpha_{\mathbf{w}}(a) = \lambda \frac{n(a, \mathbf{a})}{|\mathbf{a}|} + (1-\lambda) \frac{\sum_{\mathbf{a}} n(a, \mathbf{a})}{\sum_{\mathbf{a}} |\mathbf{a}|}$$

$$\beta_{\mathbf{w}}(b) = \lambda \frac{n(b, \mathbf{b})}{|\mathbf{b}|} + (1-\lambda) \frac{\sum_{\mathbf{b}} n(b, \mathbf{b})}{\sum_{\mathbf{b}} |\mathbf{b}|} \quad (5.19)$$

Here $n(a, \mathbf{a})$ is the number of times the word a occurs in Arabic document $\mathbf{a}$ and $|\mathbf{a}|$ is the length of that document. The summation involving $\mathbf{a}$ goes over all Arabic training strings. Bulgarian definitions are identical.

Parameter estimation with a dictionary

Parallel and comparable corpora are often difficult to come by. Translation is a rather expensive process, and we may not be able to find a parallel corpus for the pair of languages that we are dealing with. For example, we are not aware of the existence of any freely-available Arabic / Bulgarian parallel corpora. How can we estimate the paired unigrams $\{\alpha_{\mathbf{w}}, \beta_{\mathbf{w}}\}$ in this case?

If a parallel corpus is not available, we may have access to a statistical dictionary. A statistical dictionary is a matrix $T(a, b)$, which gives a likelihood that Arabic word a would be translated into Bulgarian word b. Any existing dictionary can be turned into a statistical dictionary by uniformly distributing the translation mass over all Bulgarian words b listed as translations of a. In this case, given an Arabic distribution $\alpha_{\mathbf{w}}(\cdot)$ we can define the corresponding Bulgarian distribution as follows:

$$\beta_{\mathbf{w}}(b) = \sum_{a \in \mathcal{V}_A} \alpha_{\mathbf{w}}(a) \cdot T(a, b) \quad (5.20)$$

Equation (5.20) forms the basis of the *translation* approach to ad-hoc [11] and cross-language retrieval [153, 154]. In this case, we will take the available collection of Arabic documents $\mathcal{C}$ as our training collection $\mathcal{C}_{train}$. For each Arabic document $\mathbf{w}$ we will define the Arabic distribution $\alpha_{\mathbf{w}}(\cdot)$ according to equation (5.19), and the corresponding Bulgarian distribution $\beta_{\mathbf{w}}(\cdot)$ according to equation (5.20). The resulting collection of paired distributions $\{\alpha_{\mathbf{w}}, \beta_{\mathbf{w}}\}$ will be used in equation (5.18).

Cross-lingual relevance model

With the estimation procedures in place, we can use the joint distribution defined in equation (5.18) to bridge the language gap between the query and the documents in the collection. We will use the Bulgarian query $\mathbf{q}$ to construct a relevance model over the Arabic vocabulary. Our construction follows the template established in section 3.5.3. The relevance model $RM_{\mathbf{q}}$ will assign the following probability to an Arabic word a:

$$RM_{\mathbf{q}}(a) = \frac{\int_{\mathbb{P}_{\mathcal{V}_A}} \int_{\mathbb{P}_{\mathcal{V}_B}} \alpha(a) \{\prod_{i=1}^{m} \beta(q_i)\} p_{ker,\delta}(d\alpha \times d\beta)}{\int_{\mathbb{P}_{\mathcal{V}_A}} \int_{\mathbb{P}_{\mathcal{V}_B}} \{\prod_{i=1}^{m} \beta(q_i)\} p_{ker,\delta}(d\alpha \times d\beta)}$$
$$= \frac{\sum_{\mathbf{w}} \alpha_{\mathbf{w}}(a) \prod_{i=1}^{m} \beta_{\mathbf{w}}(q_i)}{\sum_{\mathbf{w}} \prod_{i=1}^{m} \beta_{\mathbf{w}}(q_i)} \quad (5.21)$$

The summation $\sum_{\mathbf{w}}$ goes over all paired strings $\mathbf{w} = \{\mathbf{a}, \mathbf{b}\}$ in our training collection. Recall that the training collection can correspond either to a parallel corpus, or to the Arabic collection $\mathcal{C}$ augmented with a statistical dictionary. Equation (5.21) works equally well for both cases.

Document ranking

We will use negative cross-entropy over the Arabic vocabulary for ranking the documents in response to the user's query. Recall that cross-entropy $H(RM_{\mathbf{q}}||u_{\mathbf{d}})$ is rank-equivalent to the Kullback-Leiblar divergence, which forms a basis for testing the hypothesis that $\mathbf{q}$ and $\mathbf{d}$ were drawn from a common population. The cross-entropy score is defined as:

$$-H(RM_{\mathbf{q}}||u_{\mathbf{d}}) = \sum_{a \in \mathcal{V}_A} RM_{\mathbf{q}}(a) \log u_{\mathbf{d}}(a) \quad (5.22)$$

The summation above is over the Arabic words a, the relevance model $RM_{\mathbf{q}}$ is defined according to equation 5.21, and $u_{\mathbf{d}}(\cdot)$ is the frequency-based distribution of words in the Arabic document $\mathbf{d}$, defined according to equation (5.19).

5.3.2 Example of a cross-lingual relevance model

In Figure 5.3 we provide an example of a cross-lingual relevance model. In this case we started with an English query and used the method described above to estimate a relevance model over the Chinese vocabulary. We used query number 58 from the cross-language retrieval task of TREC-9. The English query was: *"environmental protection laws"*. We show 20 tokens with highest probability under the model. It is evident that many stop-words and punctuation characters are assigned high probabilities. This is not surprising, since these characters were not removed during pre-processing, and we naturally expect these characters to occur frequently in the documents that discuss any topic. However, the model also assigns high probabilities to words that one would consider strongly related to the topic of environmental protection. Note that we did not perform any morphological analysis of Chinese and did not remove any stop-words.

Q = "environmental protection laws" 环境保护法

P(word|Q)	word	meaning
0.061	，	[punctuation]
0.036	的	[possessive suffix]
0.027	。	[punctuation]
0.017	和	and
0.016	、	[punctuation]
0.009	环境	environment
0.009	了	[end of sentence]
0.008	海洋	sea
0.008	法	law
0.008	资源	resource
0.007	全国	whole country
0.007	在	in
0.006	保护	protect
0.006	污染	pollution
0.006	胶	rubber
0.006	发泡	defects in plastic
0.005	与	and
0.005	中国	china
0.005	产品	product
0.005	法律	law

Fig. 5.3. Example of a cross-lingual relevance model, estimated from query number 58 of the CLIR task of TREC-9. Shown are the 20 tokens with highest probabilities under the cross-lingual relevance model.

5.3.3 Experiments

The problem of cross-language information retrieval mirrors the problem of adhoc retrieval with one important distinction: the query $\mathbf{q} = q_1 \ldots q_k$ is given in a language that is different from the collection of documents $\mathcal{C}$. In the previous section we discussed how we can use a parallel corpus or a statistical dictionary to estimate the relevance model $RM_{\mathbf{q}}(\cdot)$ in the target language. Once a relevance model is computed, we can rank the documents using equation (5.22). In this section we carry out an evaluation of this approach on the cross-language (English - Chinese) retrieval task of TREC9.

Chinese resources

All of our cross-language experiments were performed on the dataset used in the TREC9 cross-lingual evaluation. The dataset consists of 127,938 Chinese documents, totaling around 100 million characters. We used the official set of 25 queries. We used two query representations: *short* queries used only the title field, while *long* queries used the title, description and narrative fields.

Experiments involving a bilingual dictionary used the statistical lexicon created by Xu et.al [154]. The lexicon was assembled from three parts: the

LDC dictionary, the CETA dictionary, and the statistical dictionary, learned from the Hong-Kong News corpus by applying the GIZA machine translation toolkit. Table 5.7 provides a summary of the dictionary components.

	LDC	CETA	HK News	Combined
English terms	86,000	35,000	21,000	104,997
Chinese terms	137,000	202,000	75,000	305,103

Table 5.7. Composition of the BBN bilingual lexicon

	HK News	TDT	HK News + TDT
Document pairs	18,147	46,692	64,839
English terms	28,806	67,542	83,152
Chinese terms	49,218	111,547	132,453

Table 5.8. Composition of the parallel corpus used in our experiments.

In the experiments that made use of the parallel corpus, we used the Hong-Kong News parallel dataset, which contains 18,147 news stories in English and Chinese. Because it is so small, the Hong-Kong parallel corpus has a significant word coverage problem. In order to alleviate the problem, we augmented the corpus with the TDT2 and TDT3 [24] pseudo-parallel datasets. These corpora contain 46,692 Chinese news stories along with their SYSTRAN translations into English. Since the documents are translated by software, we do not expect the quality of the TDT corpora to be as high as Hong-Kong News. We discuss the impact of adding the TDT corpora in section 5.3.3. The composition of the parallel corpus is detailed in Table 5.8.

Chinese pre-processing

The pre-processing performed on the Chinese part of the corpus was very crude, due to our limited knowledge of the language. The entire dataset, along with the Chinese queries was converted into the simplified encoding (GB). We carried out separate experiments with three forms of tokenization: (i) single Chinese characters (unigrams), (ii) half-overlapping adjacent pairs of Chinese characters (bigrams), and (iii) Chinese "words", obtained by running a simple dictionary-based segmenter, developed by F. F. Feng at the University of Massachusetts Amherst. In the following sections we will report separate figures for all three forms of tokenization, as well as a linear combination of them. We did not remove any stop-words, or any punctuation characters from either Chinese documents or queries. This results in some spurious matches and also in these characters figuring prominently in the relevance models we constructed.

System descriptions

We compare retrieval performance of the following models:

- **LM Mono-lingual baseline.** We use the basic language modeling system, which was reported as a baseline in a number of recent publications [154, 146]. The system is identical to the **LM** system described in section 5.1.3. Chinese documents D are ranked according to the probability that a Chinese query $c_1 \dots c_k$ was generated from the language model of document D.
- **RM Mono-lingual Relevance Model.** This system is included as an alternative mono-lingual baseline, and to demonstrate the degree to which relevance models degrade, when estimated in a cross-lingual setting. The system is identical to the **RM** system from section 5.1.3. Given a Chinese query $c_1 \dots c_k$, we compute the relevance model as in the ad-hoc scenario. We use relative entropy as the ranking function.
- **TM Probabilistic Translation Model.** As a cross-lingual baseline, we report the performance of our implementation of the probabilistic translation model. The model was originally proposed by Berger and Lafferty [11] and Hiemstra and de Jong [57]. We used the formulation advocated by Xu et al. [154]. We used the same statistical lexicon and the same system parameters that were reported in [154].
- **pRM Cross-lingual Relevance Model (parallel).** Given an English query $e_1 \dots e_k$, we estimate a relevance model in Chinese using techniques suggested in section 5.3.1. For probability estimation we use the combined parallel corpus (see Table 5.8). The Chinese documents are then ranked by their relative entropy with respect to the relevance model.
- **dRM Cross-lingual Relevance Model (dictionary).** We estimate the cross-lingual relevance model as suggested in section 5.3.1 Equation (5.20) is used to compute the probability of an English word e_i given a Chinese document $\mathbf{d}$. We use the lexicon reported in [154] for translation probabilities. Relative entropy is used as the ranking method.

In all cases we performed separate experiments on the three representations of Chinese: unigrams, bigrams and automatically-segmented "words". The smoothing parameters λ were tuned separately for each representation as we found that smoothing affects unigrams and bigrams very differently. The results from the three representations were then linearly combined. The weights attached to each representation were set separately for every model, in order to show best results. As an exception, the Probabilistic Translation Model was evaluated on the same representation that was used by Xu et.al.[154]. Due to the absence of the training corpus, the tuning of all parameters was performed on the testing data using a brute-force hill-climbing approach. The small number of queries in the testing dataset precluded the use of any statistical significance tests.

Baseline results

Table 5.9 shows the retrieval performance, of the described models on the TREC9 cross-language retrieval task. We use non-interpolated average precision as a performance measure. Percentage numbers indicate the difference from the mono-lingual baseline. We show results for both short and long versions of the queries. Our monolingual results form a strong baseline, competitive with the results reported by [146, 154]. This is somewhat surprising, since our processing of Chinese queries was very simplistic, and a lot of spurious matches were caused by punctuation and stop-words in the queries. We attribute the strong performance to the careful selection of smoothing parameters and combination of multiple representations. The monolingual relevance model provides an even higher baseline for both short and long queries.

The Probabilistic Translation Model achieves around 85% - 90% percent of the mono-lingual baseline. Xu et al. in [154] report the performance of the same model to be somewhat higher than our implementation (0.3100 for long queries), but still substantially lower than performance that can be achieved with relevance models. We attribute the differences to the different form of pre-processing used by Xu et al, since we used the same bilingual lexicon and the same model parameters as [154].

Cross-lingual relevance model results

	Unigrams		Bigrams		Words		Combination	
Short Queries								
LM (mono)	0.244		0.237		0.260		0.287	
RM (mono)	0.240	-2%	0.252	+6%	0.307	+18%	0.310	+8%
TM	—		—		—		0.254	-11%
dRM (dictionary)	0.277	+13%	0.268	+13%	0.277	+7%	0.280	-2%
pRM (parallel)	0.225	-8%	0.251	+6%	0.249	-4%	0.267	-7%
Long Queries								
LM (mono)	0.275		0.309		0.283		0.330	
RM (mono)	0.283	+3%	0.341	+10%	0.324	+14%	0.376	+14%
TM	—		—		—		0.276	-16%
dRM (dictionary)	0.300	+9%	0.307	-1%	0.317	+12%	0.318	-4%

Table 5.9. Average Precision on the TREC9 cross-language retrieval task. Cross-lingual relevance models perform around 95% of the strong mono-lingual baseline

Table 5.9 shows that cross-lingual relevance models perform very well, achieving 93% - 98% of the mono-lingual baseline on the combined representation. This performance is better than most previously-reported results [146, 154], which is somewhat surprising, given our poor pre-processing of Chinese. Our model noticeably outperforms the Probabilistic Translation Model on both long and short queries (see figure 5.4). It is also encouraging to see that cross-lingual relevance models perform very well on different representations of Chinese, even though they do not gain as much from the combination as the baselines.

Importance of coverage

Note that relevance models estimated using a bilingual lexicon perform better than the models estimated from the parallel corpus. We believe this is due to the fact that our parallel corpus has an acute coverage problem. The bilingual dictionary we used covers a significantly larger number of both English and Chinese words. In addition, two thirds of our parallel corpus was obtained using automatic machine translation software, which uses a limited vocabulary. It is also worth noting that the remaining part of our parallel corpus, Hong-Kong News, was also used by Xu et al. [154] in the construction of their bilingual dictionary. Refer to Table 5.8 for details.

Table 5.11 illustrates just how serious the coverage problem is. We show performance of Relevance Models estimated using just the Hong-Kong News portion of the corpus, versus performance with the full corpus. We observe tremendous improvements of over 100% which were achieved by adding the TDT data, even though this data was automatically generated using the SYSTRAN machine translation engine.

High-precision performance

Prec.	LM	RM	TM	dRM	pRM
5 docs	0.288	0.336 +17%	0.256 -11%	0.320 +11%	0.352 +22%
10 docs	0.260	0.288 +11%	0.212 -19%	0.232 -11%	0.288 +11%
15 docs	0.224	0.242 +8%	0.186 -17%	0.208 -7%	0.245 +10%
20 docs	0.212	0.222 +5%	0.170 -20%	0.196 -8%	0.208 -2%
30 docs	0.186	0.190 +2%	0.144 -23%	0.169 -9%	0.182 -2%

Table 5.10. Initial precision on the TREC9 CLIR task. Cross-lingual Relevance Models noticeably outperform the mono-lingual baselines.

Average precision is one of the most frequently reported metrics in cross-language retrieval. This metric is excellent for research purposes, but it is also important to consider user-oriented metrics. Table 5.10 shows precision at different ranks in the ranked list of documents. Precision at 5 or 10 documents is what affects users in a typical web-search setting. We observe that Cross-lingual Relevance Models exhibit exceptionally good performance in this high-precision area. Models estimated using the parallel corpus are particularly impressive, outperforming the mono-lingual baseline by 20% at 5 retrieved

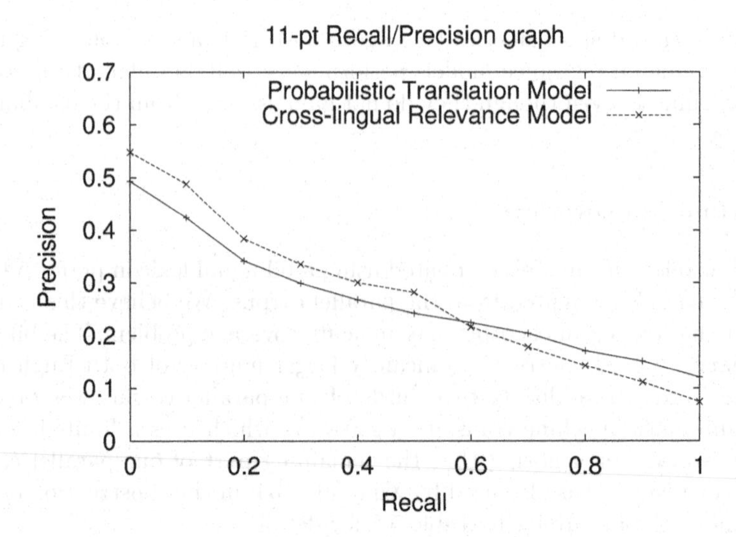

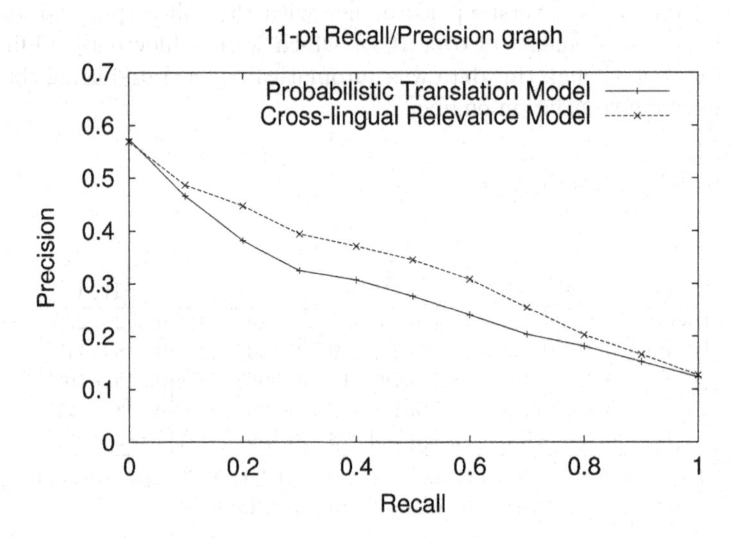

Fig. 5.4. Cross-lingual relevance models outperform the Probabilistic Translation Model on both the short (top) and long (bottom) queries.

documents. Models estimated from the bilingual dictionary perform somewhat worse, though still outperforming mono-lingual performance at 5 documents. Both estimation methods outperform the Probabilistic Translation Model. We consider these results to be extremely encouraging, since they suggest that Cross-lingual Relevance Models perform very well in the important high-precision area.

Average Precision	HK News	HK News + TDT	
Unigrams	0.1070	0.2258	+111%
Bigrams	0.1130	0.2519	+123%
"Words"	0.1210	0.2493	+106%

Table 5.11. Parallel corpus size has a very significant effect on the quality of Cross-lingual Relevance Models. Adding the pseudo-parallel TDT corpus more than doubled the average precision.

5.3.4 Significance of the cross-language scenario

The cross-language retrieval scenario discussed in the present section serves a very important purpose in the development of our model. We have demonstrated that our model can discover dependencies between words in two different languages. The generative approach allows us to associate a given bag of words (an English query) with a different bag of words (a Chinese document). The two sets of words can come from completely different vocabularies, as long as we have a parallel corpus to help us learn the coupling between them. This capacity opens a fascinating array of possible applications for our model. Our method should be able to associate any two sets of variables, as long as two conditions hold: **(i)** we have appropriate training data and **(ii)** variables in question look like words in some language.

In the following sections we will discuss three multimedia applications which extend the cross-language scenario. In each case our goal will be to associate English words with some non-linguistic features of the data. We will transform these features into something that looks like words and then use exactly the same techniques that proved to be so successful in the current section.

5.4 Handwriting Retrieval

Our first multimedia application concerns retrieval of manuscripts. A surprisingly large number of historical documents is available in handwritten form. Some of these documents are political writings, like the letters of George Washington and Thomas Jefferson, other documents represent scientific works, such as the works of Isaac Newton. These manuscripts are often available electronically, but in a form of an image rather than searchable text. Finding an item of interest in a collection of manuscripts is a tedious task because these collections are often quite large, and transcripts are not always available for every document. Even when a transcript is available, it is often of little use because there may not be a clear alignment between the transcript and the scanned pages. The alignment is particularly important to historians who prefer to work with the original pages of the manuscript, rather than with the transcript. For example, if our historian was looking for all letters written by George Washington regarding the deployment of troops near Fort Cumberland, he or she would not have an easy way of fetching images that contain the words she is interested in finding. This scenario gives rise to the problem of handwriting retrieval, which will be addressed in the present section.

Before we proceed any further we must clear away a common misconception. The handwriting retrieval task cannot be trivially solved through Optical Character Recognition (OCR). The level of noise in historical manuscripts is too high for successful application of OCR. For example, if we apply an off-the-shelf OCR system to a manuscript, we would expect to get complete garbage as output. Commercial OCR depends on being able to separate words into distinct characters, which is not possible with handwriting. If instead of commercial OCR packages we were to use a prototype handwriting recognition system (e.g. [78]), we could expect the resulting transcripts to exhibit word error rate in the range of 60-70%.[1] This means that two out of three words in the transcript would be wrong. There are a number of reasons for such poor performance. In some cases the original manuscripts are highly degraded. Often pages are scanned from microfilm rather than the original manuscript; re-scanning the original is not an option for fear of further degradation of a valuable historic article. Finally, the fluid cursive of historical manuscripts is substantially more challenging than the present day handwriting, which contains many printed characters.

5.4.1 Definition

A manuscript archive is a collection of images, each representing a complete page of handwriting. Our goal is to be able to retrieve sections of these pages in response to a text query. This task is quite different from a transcription task.

The first step in dealing with a scanned page is to identify the regions containing individual words. This is not a trivial task, but there exist a number of successful techniques, for details see [84, 107, 78]. We are going to assume that the pages have already been segmented, so we are faced with a collection of small images, each image containing a single handwritten word. An example

[1] Higher accuracy can be achieved in live handwriting recognition, where the system has access to pen stroke, position and velocity.

of such an image is shown in the left portion of Figure 5.5. Naturally, each word image corresponds to a certain word from an English vocabulary, the example in Figure 5.5 is the word *"Alexandria"*. The connection between the image and the word is trivial for a human observer, but, as we argued previously, poses a significant algorithmic challenge. In the following sections we will discuss a system that is capable of retrieving a bitmap image shown in Figure 5.5 in response to the textual query *"Alexandria"*.

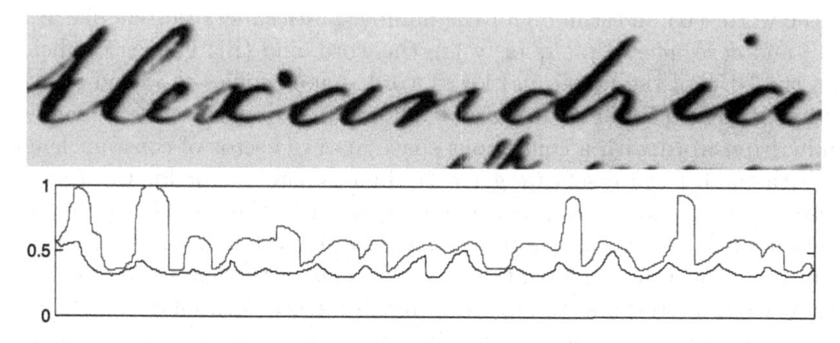

Fig. 5.5. Top: a sample handwritten word from the collection of George Washington's manuscripts. Bottom: de-slanted upper and lower profiles of the word on the left. The profiles serve as a basis for the discrete features in our algorithm.

5.4.2 Representation

We have a collection $\mathcal{C}$ of pre-segmented word images. Using the terminology from previous scenarios, each bitmap is a document. A query, in the simplest case, is a single English word. The goal is to retrieve documents (bitmaps) that would be recognized by the user as handwritten renditions of the query.

We assume that the latent, unobserved representation X of each document contains a bitmap image $\mathbf{d}$ along with the corresponding word e from the English vocabulary $\mathcal{V}$. The query generating transform Q is trivial: it takes the pair $\{\mathbf{d}, e\}$ and returns the word e. The document transform D is considerably more complex. We will take the bitmap $\mathbf{d}$ and convert it to a set of discrete *features* $f_1 \ldots f_k$. These features are intended to convey useful information about the overall *shape* of the word; description of the features will be provided below. The important thing to note is that the features f_i are discrete, i.e. they take values in some finite set $\mathcal{F}$, which can be thought of as the feature *vocabulary*. This vocabulary contains items that presumably bear some semantic content, similar to the words in a normal language. If we continue this line of thinking, we will see that the feature representation $\{f_1 \ldots f_k\}$ is essentially an *utterance* in some strange language $\mathcal{F}$. This is an important observation because it allows us to relate the problem of handwriting retrieval to the problem of cross-language retrieval, which we successfully addressed in the previous section. Finding associations between the English word e and feature words $f_1 \ldots f_k$ is no different than finding associations between English queries and words in Chinese documents.

Features and discretization

The word shape features we use in this work are described in [78]. The features include: **(i)** height, width, area and aspect ratio of the bounding box containing the word, **(ii)** an estimate for the number of *ascenders* (letters like 'd', 'b' or 'l') and *descenders* ('p', 'q' or 'y') in the word, and **(iii)** Fourier coefficients from the DFT of the upper and lower word shape profiles shown in the right portion of Figure 5.5. This feature set allows us to represent each image of a handwritten word with a continuous-space feature vector of constant length.

With these feature sets we get a 26-dimensional vector for word shapes. These representations are in continuous-space, but the model requires us to represent all feature vectors in terms of a discrete vocabulary of fixed size. One possible approach to discretization [63] is to cluster the feature vectors. Each cluster would correspond to one term in the feature vocabulary $\mathcal{F}$. However, this approach is rather aggressive, since it considers words or shapes to be equal if they fall into the same cluster.

We chose a discretization method that preserves a greater level of detail, by separately binning each dimension of a feature vector. Whenever a feature value falls into a particular bin, an associated bin number is added to the discrete-space representation of the word or shape. We used two overlapping binning schemes - the first divides each feature dimension into 10 bins while the second creates an additional 9 bins shifted by half a bin size. The overlapping bins are intended to avoid the undesirable boundary effects which happen when two very close real numbers fall on two different sides of a binning boundary. After discretization, we end up with 52 discrete features for each word image. The entire vocabulary $\mathcal{F}$ contains $26 \cdot 19 = 494$ entries.

Joint probability distribution

We can define the joint probability distribution P in exactly the same way as for the cross-language retrieval scenario. The only difference is that now all sequences involved have exactly the same length: we always have 1 English word and 52 feature words. Both components will be modeled by the multinomial distributions over their respective vocabulary. Accordingly, the parameter space Θ is given by the cross product $\mathbb{P}_{\mathcal{V}} \times \mathbb{P}_{\mathcal{F}}$, where $\mathbb{P}_{\mathcal{V}}$ is the set of all distributions over the English vocabulary $\mathcal{V}$, and $\mathbb{P}_{\mathcal{F}}$ is the set of all distributions over the feature vocabulary $\mathcal{F}$. As before, we will adopt a kernel-based density p_{ker} with delta kernels (equation 4.30). The probability of a complete observation X (an image together with the transcribed word) is:

$$P(e, f_1 \ldots f_k) = \int_{\mathbb{P}_\mathcal{V}} \int_{\mathbb{P}_\mathcal{F}} \alpha(e) \left(\prod_{i=1}^{k} \beta(f_i) \right) p_{ker,\delta}(d\alpha \times d\beta)$$
$$= \frac{1}{N} \sum_{\mathbf{x}} \alpha_{\mathbf{x}}(e) \left(\prod_{i=1}^{k} \beta_{\mathbf{x}}(f_i) \right) \tag{5.23}$$

The summation goes over all examples $\mathbf{x}$ in the training set. $\alpha_{\mathbf{x}}(\cdot)$ and $\beta_{\mathbf{x}}(\cdot)$ represent the distributions over the English words ($\mathcal{V}$) and the feature words ($\mathcal{F}$) respectively. We compute the estimates for these distributions in the same way as we did for the cross-language scenario in section 5.3.1:

$$\alpha_{\mathbf{x}}(e) = \lambda 1_{e \in \mathbf{x}} + (1-\lambda) \frac{1}{N} \sum_{\mathbf{x}} 1_{e \in \mathbf{x}}$$
$$\beta_{\mathbf{x}}(f) = \lambda 1_{f \in \mathbf{x}} + (1-\lambda) \frac{1}{N} \sum_{\mathbf{x}} 1_{f \in \mathbf{x}} \tag{5.24}$$

Here $1_{x \in \mathbf{x}}$ represents a Boolean indicator function which takes the value of one if x can be found in the representation $\mathbf{x}$ and zero otherwise. We use $1_{x \in \mathbf{x}}$ in place of the counts $n(x, \mathbf{x})$ because by definition each training example contains only one word e, and no feature value is present more than once.

Relevance model and page retrieval

We will describe two retrieval scenarios. In the first, we are only interested in retrieving single-word images, it is useful if a hypothetical historian wants to pinpoint all locations of a given word, such as *"Alexandria"*. Recall that observable documents $\mathbf{d}$ in our collection contain only the features $f_1 \ldots f_k$, they are not annotated with the corresponding word e. However, we can use equation (5.23) to come up with the probability that e would be used as a transcription, conditioned on observing the word shape features $f_1 \ldots f_k$:

$$RM_{\mathbf{d}}(e) = P(e|f_1 \ldots f_k) = \frac{\sum_{\mathbf{w}} \alpha_{\mathbf{x}}(e) \prod_{i=1}^{k} \beta_{\mathbf{x}}(f_i)}{\sum_{\mathbf{x}} \prod_{i=1}^{k} \beta_{\mathbf{x}}(f_i)} \tag{5.25}$$

Equation (5.25) allows us to perform two important tasks:

- we can use it to *annotate* a word image $\mathbf{d}$ with likely English transcription by picking one or more words e that yield a high value of $P(e|f_1 \ldots f_k)$.
- we can use it to *rank* different word images in response to a single-word query e; images $\mathbf{d}$ should be ranked in the order of decreasing $P(e|f_1 \ldots f_k)$.

The above scenario does not allow us to use multi-word queries like *"Fort Cumberland"*, which may also be quite interesting in historical analysis. To get around this inconvenience we will describe a way to rank entire passages of handwritten text. Let $\mathbf{d}^1 \ldots \mathbf{d}^n$ be a sequence of word images, representing a line or perhaps a paragraph from the manuscript. We can construct a relevance model for the entire sequence by averaging the relevance models $RM_{\mathbf{d}^i}$ that we estimate for each word image $\mathbf{d}^i$. After that we can use the classical query likelihood criterion to rank entire lines in response to a multi-word query $e_1 \ldots e_m$:

$$P(e_1 \ldots e_m | \mathbf{d}^1 \ldots \mathbf{d}^n) = \prod_{j=1}^{m} \sum_{i=1}^{n} RM_{\mathbf{d}^i}(e_j) \tag{5.26}$$

The averaging idea used above was also used in section 5.2.1 to construct a relevance model from a set of relevant examples. Our use of the query likelihood criterion was prompted by reasons of computational efficiency – with proper algorithmic optimization it is by far the fastest of all ranking methods.

5.4.3 Experiments

Experiments discussed in this section have been previously reported in [107]. We will discuss two types of evaluation. First, we briefly look at the annotation accuracy of the model described above. We train a model on a small set of hand-labeled manuscripts and evaluate how well the model was able to annotate each word in a held-out portion of the dataset. Then we turn to evaluating the model in the context of ranked retrieval.

The data set we used in training and evaluating our approach consists of 20 manually annotated pages from George Washington's handwritten letters. Segmenting this collection yields a total of 4773 images; most of these images contain exactly one word. An estimated 5-10% of the images contain segmentation errors of varying degrees: parts of words that have faded tend to get missed by the segmentation, and occasionally images contain 2 or more words or only a fragment of a word.

Evaluation methodology

Our dataset comprises 4773 total word occurrences arranged on 657 lines. Because of the relatively small size of the dataset, all of our experiments use a 10-fold randomized cross-validation, where each time the data is split into a 90% training and 10% testing sets. Splitting was performed on a line level, since we chose lines to be our retrieval unit. Prior to any experiments, the manual annotations were reduced to the root form using the Krovetz [3] morphological analyzer. This is a standard practice in information retrieval, it allows one to search for semantically similar variants of the same word. For our annotation experiments we use every word of the 4773-word vocabulary that occurs in both the training and the testing set. For retrieval experiments, we remove all function words, such as "of", "the", "and", etc. Furthermore, to simulate real queries users might pose to our system, we tested all possible combinations of 2, 3 and 4 words that occurred on the same line in the testing set. Note that these combinations are not guaranteed to co-occur anywhere in the training set. Function words were excluded from all of these combinations.

We use the standard evaluation methodology for our retrieval experiments. In response to a given query, our model produces a ranking of all lines in the testing set. Out of these lines we consider only the ones that contain all query words to be relevant. The remaining lines are assumed to be non-relevant. Then for each line in the ranked list we compute *recall* and *precision*. Recall is defined as the number of relevant lines above (and including) the current line, divided by the total number of relevant lines for the current query. Similarly, precision is defined as number of above relevant lines divided by the rank of the current line. Recall is a measure of what percent of the relevant lines we found, and precision suggests how many non-relevant lines we had to look at to achieve that recall. In our evaluation we use plots of precision vs. recall, averaged over all queries and all cross-validation repeats. We also report Mean Average Precision, which is an average of precision values at all recall points.

Results: annotating images with words

Figure 5.6 shows the performance of our model on the task of assigning word labels to handwritten images. We carried out two types of evaluation. In **position-level** evaluation, we generated a probability distribution $RM_{\mathbf{d}}(\cdot)$ for every image **d** in the testing set. Then we looked for the rank of the correct word in that distribution and averaged the resulting recall and precision over all positions. Since we did not exclude function words at this stage, position-level evaluation is strongly biased toward very common words such as "of", "the" etc. These words are generally not very interesting, so we carried out a **word-level** evaluation. Here for a given word e we look at the ranked list of all the individual word images **d** in the testing set, sorted in the decreasing order of $RM_{\mathbf{d}}(e)$. This is similar to running e as a query and retrieving all *positions* in which it could possibly occur. Recall and precision were calculated as discussed in the previous section.

From the graphs in Figure 5.6 we observe that our model performs quite well in annotation. For position-level annotation, we achieve 50% precision at rank 1, which means that for a given image **d**, half the time the word e with the highest conditional probability $RM_{\mathbf{d}}(e)$ is the correct one. Word-oriented evaluation also has close to 50% precision at rank 1, meaning that for a given word e the highest-ranked image **d** contains that word almost half the time. Mean Average Precision values are 54% and 52% for position-oriented and word-oriented evaluations respectively.

Results: retrieving images with a text query

Now we turn our attention to using our model for the task of retrieving relevant portions of manuscripts. As discussed before, we created four sets of queries: 1, 2, 3 and 4 words in length, and tested them on retrieving line

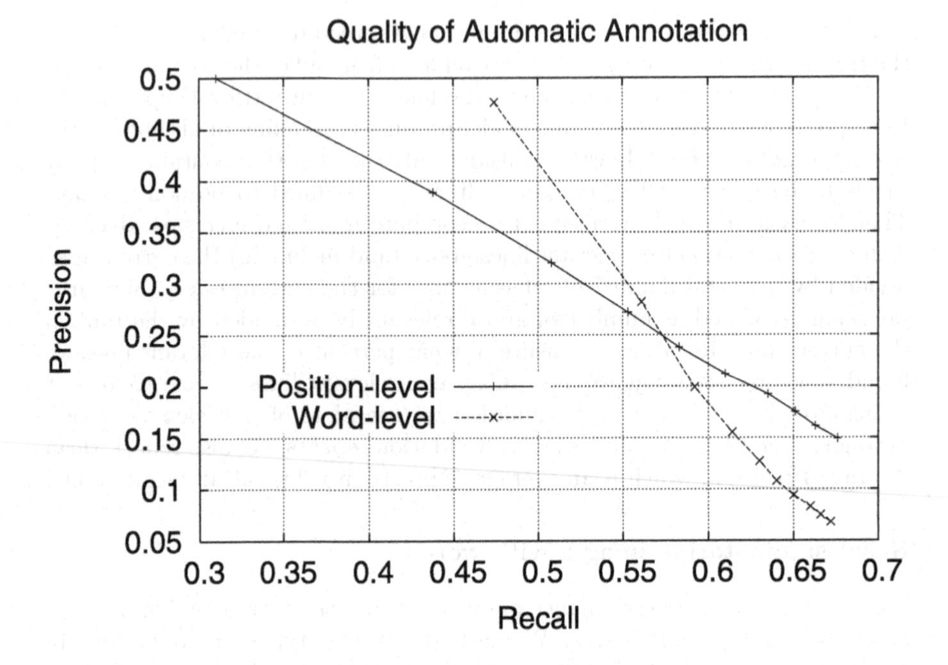

Fig. 5.6. Performance on annotating word images with words.

segments. Our experiments involve a total of 1950 single-word queries, 1939 word pairs, 1870 3-word and 1558 4-word queries over 657 lines. Figure 5.7 shows the recall-precision graphs. It is very encouraging to see that our model performs extremely well in this evaluation, reaching over 90% mean precision at rank 1. This is an exceptionally good result, showing that our model is nearly flawless when even such short queries are used. Mean average precision values were 54%, 63%, 78% and 89% for 1-, 2-, 3- and 4-word queries respectively. Figure 5.8 shows three top-ranked lines returned in response to a 4-word query *"sergeant wilper fort cumberland"*. Only one line in the entire collection is relevant to the query; the line is ranked at the top of the list.

5.5 Image Retrieval

We will now turn our attention to the problem of automatic annotation of large collections of images. This task has become particularly pertinent due to a combination of three factors: rise in the popularity and ease of digital photography, increased availability of high-bandwidth connections and the

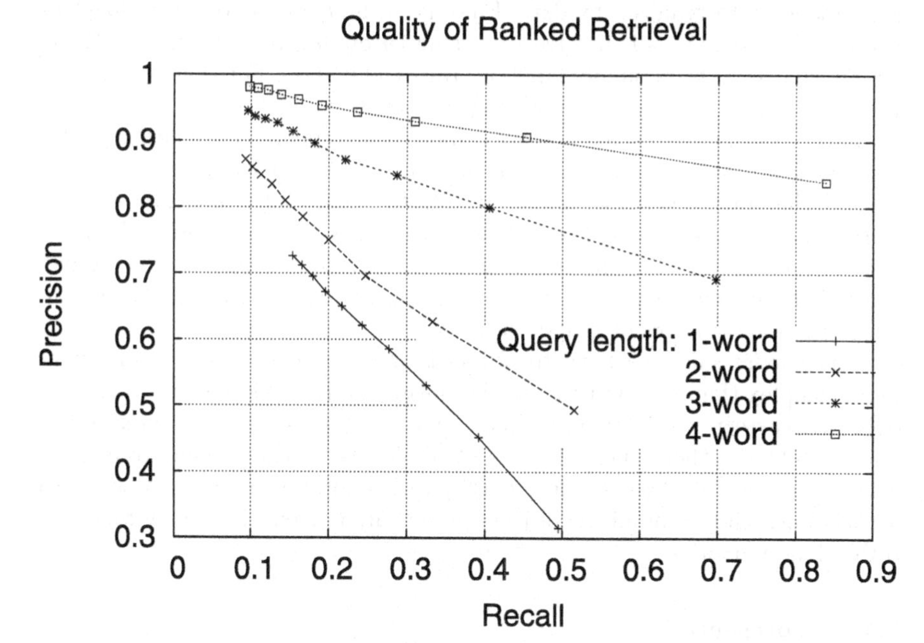

Fig. 5.7. Performance on ranked retrieval with different query sizes.

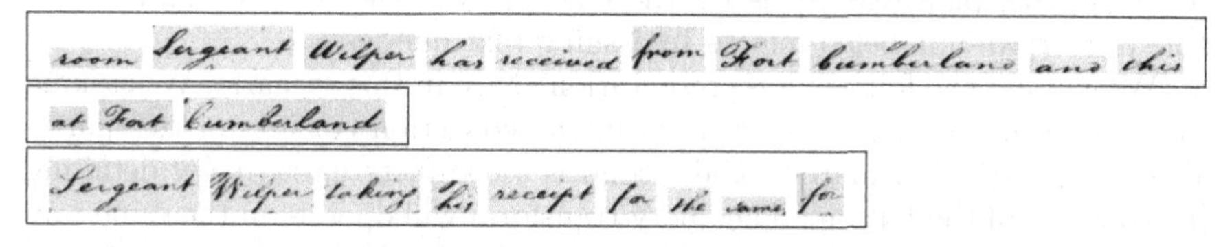

Fig. 5.8. Top 3 lines retrieved by the 4-word query: *"sergeant wilper fort cumberland"*.

ever growing storage capacity. A typical personal computer nowadays stores an astounding number of images. These may include photographs of family and friends, greeting cards, clip-art collections, diagrams and a wide variety of other images. The web, e-mail and peer-to-peer networks make it incredibly easy to share these images with wide audiences, leading to larger and larger personal collections. Unfortunately, as these collections grow they become harder and harder to organize and search through. Most of the images in personal collections are not annotated in a meaningful way: in the best case the file name contains a single keyword indicative of the contents, in the worst case images are organized by date. Finding an image in a collection like this is quite challenging, as it involves browsing through a large portion of them. An alternative is to manually label or "tag" the images, but that is a tedious endeavor that not every user is willing to consider.

In this section we discuss how we could do the labeling automatically. Our discussion will build upon previous results. In the previous section we demonstrated that our generative model is able to successfully associate English words with images showing handwritten expressions of these words. However, nothing in our model is specific to handwriting, and in this section we will show that the generative model is capable of learning relations between English words and images of general objects.

That being said, we have to admit that general images are somewhat more complex than handwritten words. In the handwriting scenario there was only one correct word for each image. We cannot assume the same about photographs or other general images. Any realistic image is going to contain a large number of objects, some perhaps more prominent than others. An accurate description of all the objects present in a given image may require a rather long narrative.

5.5.1 Representation

We have a collection $\mathcal{C}$ of un-annotated test images and another collection $\mathcal{C}_{train}$ of labeled images. Our first goal is to use the training collection to come up with plausible labels for the test images. The second goal is being able to retrieve unlabeled images by issuing text queries.

We will use the following representation space in this scenario. We assume that the latent form X of each item in the collection consists of two parts. The first part is a bitmap $\mathbf{d}$. The second is a complete narrative $\mathbf{w}$ detailing the contents of the bitmap. The query transform will operate in the same way as in the ad-hoc scenario (section 5.1.1): Q will take the latent representation $\{\mathbf{d}, \mathbf{w}\}$, discard the bitmap $\mathbf{d}$ and then return a short sub-sequence $w_1 \dots w_m$ as the query. Which words from $\mathbf{w}$ are selected as keywords will be determined by the same hypothetical mechanism we discussed in section 5.1.1.

The document generating transform D is more complex. Similar to the handwriting scenario, we will try to express the bitmap in terms of a set of feature functions. But now, since we are dealing with general images, we have to compensate for the fact that an image may contain multiple objects, and each object will be associated with a distinct type of vocabulary. The transform will consist of three distinct steps. The first step of D will be to partition the image into a set of regions $\mathbf{d}_1 \dots \mathbf{d}_k$, each containing a distinct object. The partitioning algorithm can be as simple as slicing the bitmap according to a rectangular grid. Or it can be a complex procedure like *blobworld* [22] or *normalized cuts* [124]; both attempt to find object boundaries based on local inconsistencies among the pixels.

The second step of the transform is to compute a set of representative features for each region of the bitmap. A popular set of features is suggested by [41]. Their feature set is dominated by color statistics for the region, but also includes relative position of the regions in the overall bitmap, a few shape-related features, such as area and circumference, and a number of texture statistics computed by applying Gabor filters to the region. The feature set has been used in a number of publications attempting to solve the image annotation problem. After the second step, a bitmap $\mathbf{d}$ is represented by a set of d-dimensional real-valued vectors $\mathbf{r}_1 \ldots \mathbf{r}_k$, one for each region in the bitmap.

The third step of the document transform D is optional. If we were to mimic the handwriting scenario, we would convert the real-valued feature vectors $\mathbf{r}_1 \ldots \mathbf{r}_k$ into discrete features $f_1 \ldots f_k$ taking values in some feature vocabulary $\mathcal{F}$.[2] One possible way to quantize the real-valued vectors is to *cluster* them, as was done by [41, 63]. After the clustering, each feature vector $\mathbf{r}_i$ is replaced by the number of the cluster it falls into. In this case, the feature vocabulary $\mathcal{F}$ is simply the set of cluster numbers $\{1 \ldots K\}$. After the third step, we find our representation to be identical to the representation used in the handwriting scenario: we have a set of English words $w_1 \ldots w_n$ and a set of feature words $f_1 \ldots f_k$. In this case components of the model will be defined exactly as in section 5.4.2; document ranking will be done identically to section 5.4.2. The model constructed in this fashion will be referred to as *CMRM* in the experiments.

Problems arising from clustering

The process of discretization described above involves clustering, which is a dimensionality reduction technique. As we argued in sections 4.2.7 and 4.4.3, any sort of dimensionality reduction will lead to very undesirable effects: we run a very real risk of wiping out the rare events that our user may be trying to find. For example, suppose that our database consists primarily of animal photographs, but also contains one or two technical diagrams. A typical factored model, whether it is based on clustering, mixture models or principal component analysis is very unlikely to assign a separate cluster / component to the diagrams, since there are so few of them. In all likelihood the diagrams will be folded into the general structure, i.e. collapsed into the closest animal cluster or represented as a mixture of fish, birds and insects. This is not a

[2] Unfortunately, we cannot use the quantization procedure from section 5.4.2 in the general image domain. In the handwriting domain we had a single feature vector for each bitmap, so it made sense to quantize each feature independently. In the image domain we are dealing with k regions with d features each. Since the features are supposed to be exchangeable, quantizing them individually will break the cohesion within the regions: for an image containing *grass and sky*, the texture features of the grass will end up in the same bag as the blue color features of the sky.

problem if our user is interested in animals. But it will pose a serious difficulty if he or she is seeking technical drawings.

Continuous-space generative model

As a remedy to the above problem, we could choose to skip the discretization process entirely. Instead, we could represent the bitmap by the real-valued feature vectors $\mathbf{r}_1 \ldots \mathbf{r}_k$, and try to correlate these vectors with annotation words $w_1 \ldots w_m$. The advantage of not clustering is obvious: we should be able to handle events of much finer granularity. The disadvantage is that we have to re-formulate certain parts of our representation to handle real-valued random variables. The annotation words $w_1 \ldots w_m$ will be modeled by multinomial distributions $u : \mathcal{V} \mapsto [0,1]$. The feature vectors $\mathbf{r}_1 \ldots \mathbf{r}_k$ are elements of $\mathbb{R}^d$, where d is the number of features we compute for each region. These vectors will be modeled by a probability density $h : \mathbb{R}^d \mapsto [0,1]$. Accordingly, the parameter space Θ will be the cross product $\mathbb{P}_{\mathcal{V}} \times \mathbb{F}_d$, where $\mathbb{P}_{\mathcal{V}}$ is the vocabulary simplex and $\mathbb{F}_d$ is the space of all probability density functions over $\mathbb{R}^d$. We will continue to use kernel-based density allocation, leading to the following expression for the likelihood of observing image features $\mathbf{r}_1 \ldots \mathbf{r}_k$ together with annotation $w_1 \ldots w_m$:

$$
\begin{aligned}
P(w_1 \ldots w_m, \mathbf{r}_1 \ldots \mathbf{r}_k) &= \int_{\mathbb{P}_{\mathcal{V}}} \int_{\mathcal{F}_d} \left(\prod_{i=1}^{m} u(w_i) \right) \left(\prod_{i=1}^{k} h(\mathbf{r}_i) \right) p_{ker,\delta}(du \times dh) \\
&= \frac{1}{N} \sum_{\mathbf{x}} \left(\prod_{i=1}^{m} u_{\mathbf{x}}(w_i) \right) \left(\prod_{i=1}^{k} h_{\mathbf{x}}(\mathbf{r}_i) \right) \qquad (5.27)
\end{aligned}
$$

The summation goes over all training examples $\mathbf{x} \in \mathcal{C}_{train}$.

Parameter estimation for the continuous model

As in the previous cases, all parameters in the model are tied to the individual training examples. We assume that our training collection $\mathcal{C}_{train}$ contains a set of annotated images. For each example $\mathbf{x} = \{\mathbf{d}, \mathbf{w}\}$ in the training set we need to define two components: the multinomial distribution $u_{\mathbf{x}}$ and the probability density $h_{\mathbf{x}}$. The multinomial $u_{\mathbf{x}}$ reflects the empirical distribution of words in the annotation string $\mathbf{w}$. We define this distribution in the same way as for all previous examples:

$$
u_{\mathbf{x}}(v) = \lambda \frac{n(v, \mathbf{w})}{|\mathbf{w}|} + (1-\lambda) \frac{\sum_{\mathbf{w}} n(v, \mathbf{w})}{\sum_{\mathbf{w}} |\mathbf{w}|} \qquad (5.28)
$$

As usual, $n(v, \mathbf{w})$ stands for the number of times we observe the word v in the annotation $\mathbf{w}$, the quantity $|\mathbf{w}|$ represents annotation length and the summation goes over all training annotations.

The function $h_{\mathbf{x}}$ is a probability density function over the feature space $\mathbb{R}^d$. We will use a kernel-based estimate to model $h_{\mathbf{x}}$. Recall that the bitmap portion $\mathbf{d}$ of each training example is represented by a set of k feature vectors $\mathbf{r}_1...\mathbf{r}_k$. We will place a d-dimensional Gaussian kernel on top of each vector $\mathbf{r}_j$, leading to the following probability density estimate at a given point $\mathbf{s} \in \mathbb{R}^d$:

$$h_{\mathbf{x}}(\mathbf{s}) = \frac{1}{k}\sum_{j=1}^{k} \frac{1}{\sqrt{2^d \pi^d \beta^d}} \exp\left(\frac{||\mathbf{s}-\mathbf{r}_j||}{\beta}\right) \tag{5.29}$$

Here $||\mathbf{s}-\mathbf{r}_j||$ is the Euclidean distance between feature vectors $\mathbf{s}$ and $\mathbf{r}_j$. Parameter β is the kernel *bandwidth*, it determines whether the kernel is highly peaked around its center point $\mathbf{r}_j$, or whether it is spread out.

Relevance models and image retrieval

Image annotation and ranked retrieval will be carried out in the same manner as in the handwriting scenario. For each unlabeled testing image $\mathbf{d}$ with features $\mathbf{r}_1...\mathbf{r}_k$ we will construct a relevance model $RM_{\mathbf{d}}(v)$, which represents the probability that $\mathbf{d}$ would be tagged with the word v:

$$\begin{aligned} RM_{\mathbf{d}}(v) &= P(v|\mathbf{r}_1...\mathbf{r}_k) \\ &= \frac{\int_{\mathbb{P}_{\mathcal{V}}}\int_{\mathcal{F}_d} u(v)\left\{\prod_{i=1}^{k} h(\mathbf{r}_i)\right\} p_{ker,\delta}(du \times dh)}{\int_{\mathbb{P}_{\mathcal{V}}}\int_{\mathcal{F}_d} \left\{\prod_{i=1}^{k} h(\mathbf{r}_i)\right\} p_{ker,\delta}(du \times dh)} \\ &= \frac{\sum_{\mathbf{x}} u_{\mathbf{x}}(v)\prod_{i=1}^{k} h_{\mathbf{x}}(\mathbf{r}_i)}{\sum_{\mathbf{x}}\prod_{i=1}^{k} h_{\mathbf{x}}(\mathbf{r}_i)} \end{aligned} \tag{5.30}$$

We will then use the relevance model $RM_{\mathbf{d}}(v)$ to assign most probable annotation words v to the testing image. We will also use $RM_{\mathbf{d}}(\cdot)$ to assign probabilities to multi-word queries, giving us a way to retrieve unlabeled images in response to the textual query $q_1...q_m$:

$$P(q_1...q_m|\mathbf{d}) = \prod_{j=1}^{m} RM_{\mathbf{d}}(q_j) \tag{5.31}$$

The continuous-space model described in the present section will be referred to as *CRM* in our experiments.

5.5.2 Experiments

In this section we provide a thorough evaluation of our model on two real-world tasks mentioned above. First we test the ability of our model to assign meaningful annotations to images in a held-out testing set. Then, in section 5.5.2 we evaluate the ability of our model to retrieve un-annotated images using text-only queries. In both cases we compare performance of our models to established baselines and show significant improvements over the current state-of-the-art models. Experiments discussed in this section were originally reported in [63] and [77].

Experimental Setup

To provide a meaningful comparison with previously-reported results, we use, without any modification, the dataset provided by Duygulu et al.[41][3]. This allows us to compare the performance of models in a strictly controlled manner. The dataset consists of 5,000 images from 50 Corel Stock Photo CDs. Each compact disk includes 100 images on the same topic. Each image contains an annotation of 1-5 keywords. Overall there are 371 words. Prior to modeling, every image in the dataset is pre-segmented into regions using general-purpose algorithms, such as normalized cuts [124]. We use pre-computed feature vectors for every segmented region r. The feature set consists of 36 features: 18 color features, 12 texture features and 6 shape features. For details of the features refer to [41]. We divided the dataset into 3 parts - with 4,000 training set images, 500 development set images and 500 images in the test set. The development set is used to find the smoothing meta-parameters β and λ. After fixing the parameters, we merged the 4,000 training set and 500 development set images to make a new training set. This corresponds to the training set of 4500 images and the test set of 500 images used by Duygulu et al.[41].

Results: Automatic Image Annotation

In this section we evaluate the performance of our model on the task of automatic image annotation. We are given an un-annotated image $\mathbf{d}$ and are asked to automatically produce an annotation. The automatic annotation is then compared to the held-out human annotation $\mathbf{w}$. We follow the experimental methodology used by[41, 63]. Given a set of image features $\mathbf{r}_1 \ldots \mathbf{r}_k$ we use equation (5.30) to arrive at the relevance model $RM_{\mathbf{d}}(\cdot)$. We take the top 5 words from that distribution and call them the automatic annotation of the image in question. Then, following [41], we compute annotation recall and precision for every word in the testing set. Recall is the number of images correctly annotated with a given word, divided by the number of images that have that word in the human annotation. Precision is the number of correctly annotated images divided by the total number of images annotated with that particular word (correctly or not). Recall and precision values are averaged over the set of testing words.

We compare the annotation performance of the four models: the Co-occurrence Model [93], the Translation Model [41], CMRM and the CRM.

[3] Available at http://www.cs.arizona.edu/people/kobus/research/data/eccv_2002

Models	Co-occurrence	Translation	CMRM	CRM	
#words with recall ≥ 0	19	49	66	107	+62%
Results on 49 best words, as in[63, 8]					
Mean per-word Recall	-	0.34	0.48	0.70	+46%
Mean per-word Precision	-	0.20	0.40	0.59	+48%
Results on all 260 words					
Mean per-word Recall	0.02	0.04	0.09	0.19	+111%
Mean per-word Precision	0.03	0.06	0.10	0.16	+60 %

Table 5.12. Comparing recall and precision of the four models on the task of automatic image annotation. CRM substantially outperforms all other models. Percent improvements are over the second-best model (CMRM). Both CRM and CMRM significantly outperform the state-of-the-art translation model.

We report the results on two sets of words: the subset of *49 best* words which was used by [41, 63], and the complete set of all 260 words that occur in the testing set. Table 5.12 shows the performance on both word sets. The figures clearly show that the continuous-space generative model (CRM) substantially outperforms the other models and is the only one of the four capable of producing reasonable mean recall and mean precision numbers when every word in the test set is used. In Figure 5.9 we provide sample annotations for the two best models in the table, CMRM and CRM, showing that the CRM is considerably more accurate.

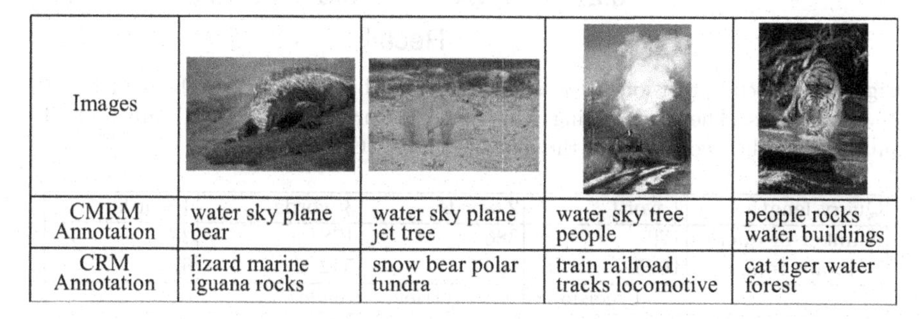

Fig. 5.9. The generative model based on continuous features (CRM) performs substantially better than the cluster-based cross-media relevance model (CMRM) for annotating images in the test set.

Results: Ranked Retrieval of Images

In this section we turn our attention to the problem of ranked retrieval of images. In the retrieval setting we are given a text query $q_1 \ldots q_m$ and a testing collection of un-annotated images. For each testing image $\mathbf{d}$ we use equation (5.30) to get the relevance model $RM_{\mathbf{d}}(\cdot)$. All images in the collection

Fig. 5.10. Example: top 5 images retrieved by CRM in response to text query *"cars track"*

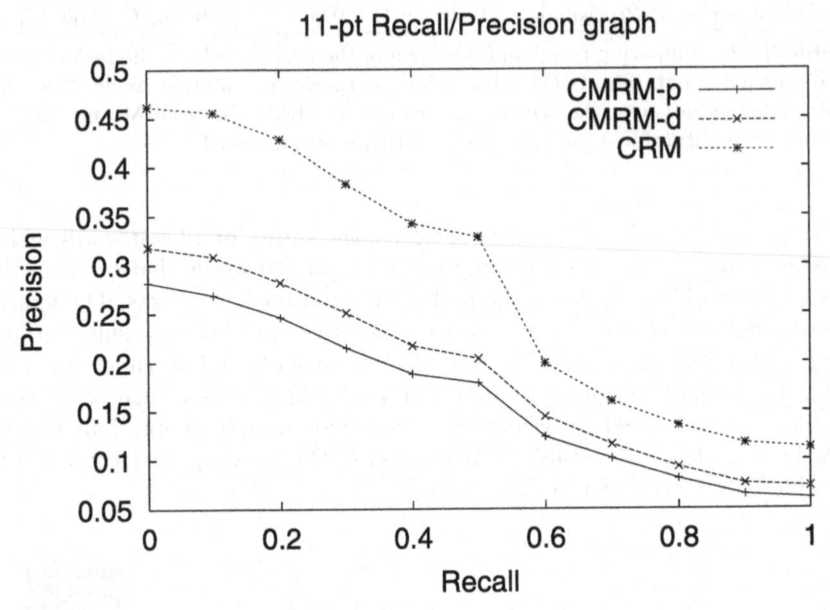

Fig. 5.11. Recall-precision curves showing retrieval performance of the models with 2-word queries. The model using real-valued feature vectors (CRM) substantially outperforms the two flavors of the discrete model (CMRM).

Query length	1 word	2 words	3 words	4 words
Number of queries	179	386	178	24
Relevant images	1675	1647	542	67
Precision after 5 retrieved images				
CMRM	0.1989	0.1306	0.1494	0.2083
CRM	0.2480 **+25%**	0.1902 **+45%**	0.1888 **+26%**	0.2333 +12%
Mean Average Precision				
CMRM	0.1697	0.1642	0.2030	0.2765
CRM	0.2353 **+39%**	0.2534 **+54%**	0.3152 **+55%**	0.4471 +61%

Table 5.13. Comparing CRM and CMRM on the task of image retrieval. Our model outperforms the CMRM model by a wide margin on all query sets. Boldface figures mark improvements that are statistically significant according to sign test with a confidence of 99% (p-value < 0.01).

are ranked according to the query likelihood criterion under $RM_{\mathbf{d}}(\cdot)$. We use four sets of queries, constructed from all 1-, 2-, 3- and 4-word combinations of words that occur at least twice in the testing set. An image is considered relevant to a given query if its *manual* annotation contains all of the query words. As our evaluation metrics we use precision at 5 retrieved images and non-interpolated average precision[4], averaged over the entire query set. Precision at 5 documents is a good measure of performance for a casual user who is interested in retrieving a couple of relevant items without looking at too much junk. Average precision is more appropriate for a professional user who wants to find a large proportion of relevant items.

Table 5.13 shows the performance of CRM on the four query sets, contrasted with performance of the CMRM baseline on the same data. We observe that our model substantially outperforms the CMRM baseline on every query set. Improvements in average precision are particularly impressive, our model outperforms the baseline by 40 - 60 percent. All improvements on 1-, 2- and 3-word queries are statistically significant based on a sign test with a p-value of 0.01. We are also very encouraged by the precision our model shows at 5 retrieved images: precision values around 0.2 suggest that an average query always has a relevant image in the top 5. Figure 5.10 shows top 5 images retrieved in response to the text query *"cars track"*.

5.6 Video Retrieval

Collections of video footage represent a mesmerizing source of information. Personal video collections are becoming more commonplace, and they are typically even more disorganized than collections of photographs. In addition, there exist countless public archives of digitized video. These archives contain anything from news reports, to movies, to historical footage, to recordings made by security cameras. A particularly fascinating example is the collection assembled by the Shoah Visual History Foundation[5]. Established by Steven Spielberg, the archive is truly immense. It contains hundreds of thousands of hours of testimony from nearly 52,000 survivors of the Holocaust. Searching through a video archive of this size is a significant challenge. Video footage has no innate annotations, so finding a certain segment typically involves the user rewinding and fast-forwarding until he sees what he likes. This is feasible on a relatively small scale, but completely out of the question if we are dealing with 100,000 hours of potentially relevant video. A typical solution is hand-labeling the tapes with meta-data. For example, the Shoah Foundation Archive provides extensive biographical information for each interviewee; it includes gender, place and date of birth, religious identity, languages spoken

[4] Average precision is the average of precision values at the ranks where relevant items occur.

[5] http://www.vhf.org

and a lot of other information. However, all this information cannot help us if we are looking for specific content on the tape. For example, we have no way of searching for segments that contain something other than the interviewee: such as photographs, letters or other articles that may have been filmed during the testimony.

One possible solution to the video retrieval problem is to use Automatic Speech Recognition on the audio accompanying the footage. However, this assumes that **(i)** everything shown in the video is appropriately narrated and **(ii)** we have a highly accurate speech recognition system for the language in which the narrative was given. The second requirement is not trivial, for instance the Shoah Archive contains testimonies in 31 languages, ranging from English and German to Croatian and Romani. Finding a speech recognizer that is able to handle a digitally-sparse language like Romani may turn out to be a challenge.

In this section we will describe our solution to the video retrieval problem. The solution involves *annotating* individual keyframes in the video with appropriate keywords. The keywords will describe the objects contained in the frame. These annotations will allow us to search through the video by issuing text queries. An important advantage of this approach is that it is language-independent; an airplane in the video footage is going to look like an airplane regardless of the narration language. Consequently, our approach will allow us to issue an English query, e.g. *"riots"*, and retrieve relevant video footage from Arabic, Chinese or Russian video collections or streams.[6]

5.6.1 Representation

Our video retrieval model will closely follow the image retrieval model described in the previous chapter. A video archive is a collection of keyframes automatically extracted from video. Each keyframe is a bitmap image, no different from the images discussed in the previous chapter. We hypothesize that in its latent form each keyframe is coupled with a long English narrative describing the objects contained in the keyframe. The representation will be identical to the image retrieval scenario (see section 5.5.1). We will segment the keyframes into a set of regions and compute the feature vectors $\mathbf{r}_1 \ldots \mathbf{r}_k$ for each region. Given the superior performance of the continuous-space model (CRM), we will not discretize the features. The generative model will take the form of equation (5.27), and the relevance model will be given by equation (5.30). Parameters will be estimated in exactly the same manner as suggested in section 5.5.1.

There will be two minor differences in the application of our model to the video domain. First, instead of using sophisticated segmentation algorithms

[6] Our current solution completely ignores the audio component of video recordings. An exciting future extension would be to incorporate the audio features, either directly or as ASR output.

like blob-world [22] and normalized-cuts [124], we will slice up each keyframe according to a rectangular grid. There are two motivations for this step: (i) the computational expense of segmentation becomes a serious problem for video and (ii) there are some indications [44, 76] that rectangular regions perform better than more sophisticated segments, at least in the datasets we experimented with.

The second difference from the previous scenario will be in the way we pick the smoothing parameter λ in equation (5.28). Previously, we treated λ as a constant independent of the annotation length $|\mathbf{w}|$. The value of the constant was determined by optimizing retrieval performance on the development set. In the current scenario we will set λ to be the fraction $\frac{|\mathbf{w}|}{K}$, where $|\mathbf{w}|$ is the annotation length of the example and K is a large constant, bigger than the maximum observed annotation length $\max_{\mathbf{w}}\{|\mathbf{w}|\}$. The effect of setting λ in this fashion will be to compensate for the varying lengths of training annotations. This is necessary because of a somewhat erratic nature of example annotations in our video collection. A video frame showing a reporter can be annotated either with a single word [*"face"*], or with a set of words [*"face"*, *"male_face"*, *"reporter"*, *"indoors"*, *"studio_setting"*]. When we keep λ constant, the first annotation would lead to a maximum likelihood probability of 1 for the word *"face"*. The second annotation would give the same word a probability of $\frac{1}{5}$. By setting $\lambda = \frac{|\mathbf{w}|}{K}$ we make the probability estimates comparable under the two examples: the probability of *"face"* in both cases will be $\frac{1}{K}$. The effect is that all annotations look like they have the same length K; we will refer to this model as the *fixed-length CRM.*

5.6.2 Experiments

Experiments discussed in this section were originally reported in [44, 76]. We provide the experimental results of the retrieval task over a keyframe dataset, which is a subset of the Video Track dataset of the TREC conference. The data set consists of 12 MPEG files, each of which is a 30-minutes video section of CNN or ABC news and advertisements. 5200 keyframes were extracted and provided by NIST for this dataset. The TREC participants annotated a portion of the videos. The word vocabulary for human annotation is structured as a hierarchical tree, where each annotation word represents a distinct node. Many keyframes are annotated hierarchically, e.g. a keyframe can be assigned a set of words like "face, female_face, female_news_person". The annotation length for keyframes can vary widely. There are 137 keywords in the whole dataset after we remove all the audio annotations.We randomly divide the dataset into a training set (1735 keyframes), a development set (1735 keyframes) and a testing set (1730 keyframes). The development set is used to find system parameters, and then merged into the training set after we find the parameters.

Every keyframe in this set is partitioned into rectangular grids, and a feature vector is then calculated for every grid region. The number of rectangles is empirically selected (using the training and development sets), it is 35 for each sample. The feature set consists of 30 features: 18 color features (including region color average, standard deviation and skewness) and 12 texture features (Gabor energy computed over 3 scales and 4 orientations).

In our retrieval experiments, we use three sets of queries, constructed from all 1-, 2-, and 3-word combinations that occur at least 10 times in the testing set. For each set of queries, we perform the comparative experiments over the two different methods for setting the smoothing parameter λ. An image is considered relevant to a given query if its unobserved manual annotation contains all the query words. As used in [44, 76, 77], evaluation metrics are precision at 5 retrieved keyframes and non-interpolated average precision, averaged over the entire query set. These two different metrics are good measures suitable for distinct needs of casual users and professional users.

Query length	1 word	2 words	3 words
Number of queries	107	431	402
Relevant images	6649	12553	11023
Mean precision after 5 retrieved keyframes			
CRM	0.36	0.33	0.42
CRM (fixed length)	0.49	0.47	0.58
Mean Average Precision			
CRM	0.26	0.19	0.25
CRM (fixed-length)	0.30	0.26	0.32

Table 5.14. Performance of CRM on the TREC video retrieval task.

Fig. 5.12. First 4 ranked results for the query *"basketball"*. Top row: retrieved by CRM. Bottom row: retrieved by fixed-length CRM.

Table 5.14 shows the details of the performance of our two different methods over the three sets of queries. We can observe that fixed-length CRM substantially outperforms the original formulation. The improvements are 15%,

Fig. 5.13. First 4 ranked results for the query *"outdoors, sky, transportation"*. Top row: CRM results. Bottom row: fixed-length CRM results.

37% and 33% on the 1-, 2- and 3-word query sets respectively. The precision at 5 retrieved keyframes also indicates that for fixed-length CRM, half of the top 5 images are relevant to the query. Figures 5.12 and 5.13 show the top 4 images in the ranked lists corresponding to the text queries *"basketball"* and *"outdoors sky transportation"* respectively. In both figures, the top row represents performance of the regular CRM, the bottom row shows the results from fixed-length CRM.

5.7 Structured search with missing data

In this section we will look at the problem of locating relevant records with missing field values. We assume that our user is searching a semi-structured database containing records, some of which will satisfy his information need. Each record is a set of natural-language fields with clearly-defined semantics for each field. For example, library database records may include fields like *author*, *subject* and *audience*. In order to locate the relevant records the user will generate structured requests that describe his information need, for example: $\mathbf{q}$ = {*subject='physics,gravity' AND audience='grades 1-4'*}.[7]

When the database is complete and every value of every record is available, such queries can be efficiently answered by a relational database system. Furthermore, using the above query as part of an SQL statement could furnish our user with a number of useful summary statistics, for instance the list of authors who write on the subject of interest, or the list of other subjects covered by the same authors that write about gravity for beginners. Unfortunately, many real-world semi-structured databases exhibit certain pathological properties that significantly complicate the use of conventional database

[7] For the purposes of this work, we will focus on simple conjunctive queries. Extending our model to more complex queries is reserved for future research.

engines. To illustrate this point we consider the *National Science Digital Library (NSDL)* database[8], which contains various educational resources such as tutorials, instructional materials, videos and research reports. The NSDL collection poses the following challenges:

1. **Missing fields.** Relational engines typically assume complete information for every record in the database. Semi-structured collections often have incomplete markup. For example 24% of the items in the NSDL database have no subject field, 30% are missing the author information, and over 96% mention no target audience (readability level). This means that a relational query for elementary school material will consider at most 4% of all potentially relevant resources in the NSDL collection.
2. **Natural language values.** Even when an item contains values for all query fields, we are still not guaranteed an easy decision of whether or not it is relevant to the user. The problem stems from the fact that field values may be expressed as free-form text, rather than a set of agreed-upon keywords from a closed vocabulary. For example, the *audience* field in the NSDL collection contains values such as *'elementary'*, *'K-4'*, *'second grade'* and *'learner'*, all of which are clearly synonymous. Deciding whether an item targeting *'K-6'* students is relevant to a query that specifies *elementary school* is not trivial.
3. **Inconsistent schemata.** In addition to ambiguous semantics inherent in natural language, the semantics of fields themselves are not always consistent. For example, the NSDL collection includes both an *audience* and an *audienceEducationLevel* fields. The meaning of these fields is often synonymous – they both typically specify the education level of readers' the given record assumes. Problems like this often arise when two or more databases with different schemata are merged into a single digital library.

The goal of this section is to introduce a retrieval model that will be capable of answering complex structured queries over a semi-structured collection that exhibits the problems mentioned above. The approach is based on the idea that plausible values for a given field could be inferred from the context provided by the other fields in the record. For instance, it is unlikely that a resource titled *'Transductive SVMs'* and containing highly technical language in will be aimed at elementary-school students. In the remainder of this section, we will describe how relevance models allow us to guess the values of unobserved fields. At the intuitive level, the relevance model will leverage the fact that records similar in one respect will often be similar in others. For example, if two resources share the same author and have similar titles, they are likely to be aimed at the same audience.

[8] http://www.nsdl.org

5.7.1 Representation of documents and queries

Each record (document) $\mathbf{d}$ in the database consists of a set of fields $\mathbf{d}^1 \ldots \mathbf{d}^m$. Each field $\mathbf{d}^i$ is a sequence of discrete variables (words) $\mathbf{d}^i_1 \ldots \mathbf{d}^i_{n_i}$, taking values in the field vocabulary $\mathcal{V}_i$. We allow each field to have its own vocabulary $\mathcal{V}_i$ since we generally do not expect author names to occur in the audience field, etc. When a record contains no information for the i'th field, we assume $n_i{=}0$ for that record. The user's query $\mathbf{q}$ takes the same representation as a document: $\mathbf{q}{=}\{\mathbf{q}^i_j{\in}\mathcal{V}_i : i{=}1..m, j = 1..n_i\}$.

5.7.2 Probability distribution over documents and queries

The random variables X^i_j comprising our representation are assummed to be fully exchangeable within each field. We will use $\mathbf{p}^i$ to denote a probability distribution over $\mathcal{V}_i$, i.e. a set of probabilities $\mathbf{p}^i(v){\in}[0,1]$, one for each word v, obeying the constraint $\Sigma_v \mathbf{p}^i(v) = 1$. The set of all possible language models over $\mathcal{V}_i$ will be denoted as the probability simplex $\mathbb{P}_i$. With these assumptions, we can represent the joint probability of observing a record $\mathbf{q}$ (document or query) as:

$$P(\mathbf{q}^{1\ldots m}_{1\ldots n_i}) = \int_{\mathbb{P}_1\ldots\mathbb{P}_m} \left[\prod_{i=1}^{m}\prod_{j=1}^{n_i} \mathbf{p}^i(\mathbf{q}^i_j)\right] p_{ker,\delta}(d\mathbf{p}^1 \times \ldots \times d\mathbf{p}^m) \quad (5.32)$$

Note the similarity of equation (5.32) to the joint probability of cross-language representations described in section 5.3. The main difference is that now we are dealing with m distinct "languages", each representing a separate field in our records. As before, we use the kernel-based generative density estimate $p_{ker,\delta}$. Specifically, we define $p_{ker,\delta}(d\mathbf{p}^1 \times \ldots \times d\mathbf{p}^m)$ to have mass $\frac{1}{N}$ when its argument $\mathbf{p}^1 \ldots \mathbf{p}^m$ corresponds to one of the N records $\mathbf{d}$ in the training part $\mathcal{C}_t$ of our collection, and zero otherwise:

$$p_{ker,\delta}(d\mathbf{p}^1 \ldots d\mathbf{p}^m) = \frac{1}{N} \sum_{\mathbf{d}\in\mathcal{C}_t} \prod_{i=1}^{m} 1_{\mathbf{p}^i = \mathbf{p}^i_{\mathbf{d}}} \quad (5.33)$$

Here 1_x is the Boolean indicator function that returns 1 when its predicate x is true and zero when it is false. $\mathbf{p}^i_{\mathbf{d}}$ is the language model associated with field i of the training record $\mathbf{d}$ and defined as follows:

$$\mathbf{p}^i_{\mathbf{d}}(v) = \frac{n(v, \mathbf{d}^i) + \mu_i c_v}{n_i + \mu_i} \quad (5.34)$$

$n(v, \mathbf{d}_i)$ represents the number of times the word v was observed in the i'th field of $\mathbf{d}$, n_i is the length of the i'th field, and c_v is the relative frequency of v in the entire collection. Meta-parameters μ_i allow us to control the amount of smoothing applied to language models of different fields; their values are set empirically on a held-out portion of the data.

5.7.3 Structured Relevance Model

Given the user's query $\mathbf{q}$, we define a relevance model $RM_{\mathbf{q}}^k$ for every field k in our records:

$$RM_{\mathbf{q}}^k(v) = \frac{P(\mathbf{q}^1 \ldots v \circ \mathbf{q}^k \ldots \mathbf{q}^m)}{P(\mathbf{q}^1 \ldots \mathbf{q}^i \ldots \mathbf{q}^m)} \tag{5.35}$$

We use $v \circ \mathbf{q}^k$ to denote appending word v to the string $\mathbf{q}^k$. Both the numerator and the denominator are computed using equation (5.32), which takes the following form after we substitute the kernel-based density given by equation (5.33):

$$\begin{aligned} RM_{\mathbf{q}}^k(v) &= \frac{\int_{\mathbb{P}_1 \ldots \mathbb{P}_m} \left[\mathbf{p}^k(v) \prod_{i=1}^m \prod_{j=1}^{n_i} \mathbf{p}^i(\mathbf{q}_j^i) \right] p_{ker,\delta}(d\mathbf{p}^1 \times \ldots \times d\mathbf{p}^m)}{\int_{\mathbb{P}_1 \ldots \mathbb{P}_m} \left[\prod_{i=1}^m \prod_{j=1}^{n_i} \mathbf{p}^i(\mathbf{q}_j^i) \right] p_{ker,\delta}(d\mathbf{p}^1 \times \ldots \times d\mathbf{p}^m)} \\ &= \frac{\sum_{\mathbf{d} \in \mathcal{C}_t} \mathbf{p}_{\mathbf{d}}^k(v) \prod_{i=1}^m \prod_{j=1}^{n_i} \mathbf{p}_{\mathbf{d}}^i(\mathbf{q}_j^i)}{\sum_{\mathbf{d} \in \mathcal{C}_t} \prod_{i=1}^m \prod_{j=1}^{n_i} \mathbf{p}_{\mathbf{d}}^i(\mathbf{q}_j^i)} \end{aligned} \tag{5.36}$$

The quantity $RM_{\mathbf{q}}^k(v)$ is interpreted as the likelihood that word v should appear in the k'th field of a record whose incomplete observed form is given by $\mathbf{q}^1 \ldots \mathbf{q}^m$. Note that $\mathbf{q}^k$ may be completely empty in the observed record: we will still be able to estimate a relevance model for that field, as long as one of the other fields is provided. For example, if $\mathbf{q}$ contained only the *subject* and *audience* fields, we would be able to estimate likely values for *author*, *title*, and any other field in our data. We would also estimate probability distributions over *subject* and *audience* fields, suggesting values that are statistically related to the values specified by the user.

5.7.4 Retrieving Relevant Records

Once we have computed relevance models $RM_{\mathbf{q}}^k(\cdot)$ for each of the m fields, we can rank *testing* records $\mathbf{d}$ by their similarity to these relevance models. As a similarity measure we use a weighted variant of the cross-entropy ranking criterion, which was discussed in section 3.6.2.

$$H(RM_{\mathbf{q}}^{1..m} || \mathbf{d}^{1..m}) = \sum_{k=1}^m \alpha_k \sum_{v \in \mathcal{V}_i} RM_{\mathbf{q}}^k(v) \log \mathbf{p}_{\mathbf{d}}^k(v) \tag{5.37}$$

The outer summation goes over every field of interest, while the inner extends over all the words in the vocabulary of the k'th field. $RM_{\mathbf{q}}^k(v)$ are computed according to equation (5.36), while $\mathbf{p}_{\mathbf{d}}^k$ are estimated from equation (5.34). Meta-parameters α_k allow us to vary the importance of different fields in the final ranking; the values are selected on a held-out portion of the data.

5.7.5 Experiments

Our experiments closely follow those performed for the previous retrieval scenarios. We note that this task is not a typical IR task because the fielded structure of the query is a critical aspect of the processing, not one that is largely ignored in favor of pure content based retrieval. On the other hand, the approach used is different from most DB work because cross-field dependencies are a key component of the technique. In addition, the task is unusual for both DB and IR communities because it considers an uncommon case where the fields in the query do not occur at all in the documents being searched. The experiments reported in this section were originally reported in [79], for additional details see [155].

Dataset and queries

	records covered	average length	unique values
title	655,673 (99%)	7	102,772
description	514,092 (78%)	38	189,136
subject	504,054 (77%)	12	37,385
content	91,779 (14%)	743	575,958
audience	22,963 (3.5%)	4	119

Table 5.15. Summary statistics for the five NSDL fields used in our retrieval experiments.

We tested the performance of our model on a January 2005 snapshot of the National Science Digital Library repository. The snapshot contains a total of 656,992 records, spanning 92 distinct (though sometimes related) fields. Only 7 of these fields are present in every record, and half the fields are present in less than 1% of the records. An average record contains only 17 of the 92 fields. Our experiments focus on a subset of 5 fields (*title, description, subject, content* and *audience*). These fields were selected for two reasons: (i) they occur frequently enough to allow a meaningful evaluation and (ii) they seem plausible to be included in a potential query.[9] Of these fields, *title* represents the title of the resource, *description* is a very brief abstract, *content* is a more detailed description (but not the full content) of the resource, *subject* is a library-like classification of the topic covered by the resource, and *audience* reflects the target reading level (e.g. *elementary school* or *post-graduate*). Summary statistics for these fields are provided in Table 5.15.

The dataset was randomly split into three subsets: the **training** set, which comprised 50% of the records and was used for estimating the relevance models

[9] The most frequent NSDL fields (*id, icon, url, link* and four distinct *brand* fields) seem unlikely to be used in user queries.

as described above; the **held-out** set, which comprised 25% of the data and was used to tune the smoothing parameters μ_i and the cross-entropy weights α_i; and the **evaluation** set, which contained 25% of the records and was used to evaluate the performance of the tuned model.

Our experiments are based on a set of 127 automatically generated queries. We randomly split the queries into two groups, 64 for training and 63 for evaluation. The queries were constructed by combining two randomly picked *subject* words with two *audience* words, and then discarding any combination that had less than 10 exact matches in any of the three subsets of our collection. This procedure yields queries such as $\mathbf{q}_{91}$={*subject:'artificial intelligence' AND audience='researchers'*}, or bq_{101}={*subject:'philosophy' AND audience='high school'*}.

Evaluation paradigm

We evaluate our model by its ability to find "relevant" records in the face of missing values. We define a record **d** to be relevant to the user's query **q** if every keyword in **q** is found in the corresponding field of **d**. For example, in order to be relevant to Q_{101} a record must contain the word *'philosophy'* in the subject field and words *'high'* and *'school'* in the audience field. If either of the keywords is missing, the record is considered non-relevant. This definition of relevance is unduly conservative by the standards of Information Retrieval researchers. Many records that might be considered relevant by a human annotator will be treated as non-relevant, artificially decreasing the accuracy of any retrieval algorithm. However, our approach has the advantage of being fully automatic: it allows us to test our model on a scale that would be prohibitively expensive with manual relevance judgments.

When the testing records are fully observable, achieving perfect retrieval accuracy is trivial: we simply return all records that match all query keywords in the subject and audience fields. As we stated earlier, our main interest concerns the scenario when parts of the testing data are missing. We are going to simulate this scenario in a rather extreme manner by *completely* removing the *subject* and *audience* fields from all testing records. This means that a straightforward approach – matching query fields against record fields – will yield no relevant results. Our approach will rank testing records by comparing their *title, description* and *content* fields against the query-based relevance models, as discussed above.

We will use the standard rank-based evaluation metrics: *precision* and *recall.* Let N_R be the total number of records relevant to a given query, suppose that the first K records in our ranking contain N_K relevant ones. Precision at rank K is defined as $\frac{N_K}{K}$ and recall is defined as $\frac{N_K}{N_R}$. Average precision is defined as the mean precision over all ranks where relevant items occur. R-precision is defined as precision at rank $K{=}N_R$.

Baseline systems

Our experiments will compare the ranking performance of the following retrieval systems:

cLM is a *cheating* version of un-structured text search using a state-of-the-art language-modeling approach [106]. We disregard the structure, take all query keywords and run them against a *concatenation* of all fields in the testing records. This is a "cheating" baseline, since the concatenation includes the *audience* and *subject* fields, which are supposed to be missing from the testing records. We use Dirichlet smoothing [68], with parameters optimized on the training data. This baseline mimics the core search capability currently available on the NSDL website.

bLM is a combination of SQL-like structured matching and unstructured search with query expansion. We take all training records that contain an exact match to our query and select 10 highly-weighted words from the *title*, *description*, and *content* fields of these records. We run the resulting 30 words as a language modeling query against the concatenation of *title*, *description*, and *content* fields in the testing records. This is a non-cheating baseline.

bMatch is a structured extension of bLM. As in bLM, we pick training records that contain an exact match to the query fields. Then we match 10 highly-weighted *title* words, against the *title* field of testing records, do the same for the *description* and *content* fields, and merge the three resulting ranked lists. This is a non-cheating baseline that is similar to our model (SRM). The main difference is that this approach uses exact matching to select the training records, whereas SRM leverages a best-match language modeling algorithm.

SRM is the Structured Relevance Model, as described above.

Note that our baselines do not include a standard SQL approach directly on testing records. Such an approach would have perfect performance in a "cheating" scenario with observable *subject* and *audience* fields, but would not match any records when the fields are removed.

Experimental results

Table 5.16 shows the performance of our model (SRM) against the three baselines. The model parameters were tuned using the 64 training queries on the *training* and *held-out* sets. The results are for the 63 test queries run against the *evaluation* corpus. (Similar results occur if the 64 training queries are run against the *evalution* corpus.)

The upper half of Table 5.16 shows precision at fixed recall levels; the lower half shows precision at different ranks. The *%change* column shows relative difference between our model and the baseline bLM. The *improved* column

	cLM	bMatch	bLM	SRM	%change	improved
Rel-ret:	949	582	914	861	-5.80	26/50
Interpolated Recall - Precision:						
at 0.00	0.3852	0.3730	0.4153	0.5448	31.2	**33/49**
at 0.10	0.3014	0.3020	0.3314	0.4783	44.3	**42/56**
at 0.20	0.2307	0.2256	0.2660	0.3641	36.9	**40/59**
at 0.30	0.2105	0.1471	0.2126	0.2971	39.8	**36/58**
at 0.40	0.1880	0.1130	0.1783	0.2352	31.9	**36/58**
at 0.50	0.1803	0.0679	0.1591	0.1911	20.1	32/57
at 0.60	0.1637	0.0371	0.1242	0.1439	15.8	27/51
at 0.70	0.1513	0.0161	0.1001	0.1089	8.7	21/42
at 0.80	0.1432	0.0095	0.0901	0.0747	-17.0	18/36
at 0.90	0.1292	0.0055	0.0675	0.0518	-23.2	12/27
at 1.00	0.1154	0.0043	0.0593	0.0420	-29.2	9/23
Avg.Prec.	0.1790	0.1050	0.1668	0.2156	29.25	**43/63**
Precision at:						
5 docs	0.1651	0.2159	0.2413	0.3556	47.4	**32/43**
10 docs	0.1571	0.1651	0.2063	0.2889	40.0	**34/48**
15 docs	0.1577	0.1471	0.1841	0.2360	28.2	**32/49**
20 docs	0.1540	0.1349	0.1722	0.2024	17.5	28/47
30 docs	0.1450	0.1101	0.1492	0.1677	12.4	29/50
100 docs	0.0913	0.0465	0.0849	0.0871	2.6	**37/57**
200 docs	0.0552	0.0279	0.0539	0.0506	-6.2	33/53
500 docs	0.0264	0.0163	0.0255	0.0243	-4.5	26/48
1000 docs	0.0151	0.0092	0.0145	0.0137	-5.8	26/50
R-Prec.	0.1587	0.1204	0.1681	0.2344	39.44	**31/49**

Table 5.16. Performance of the 63 test queries on the evaluation data. Bold figures show statistically significant differences. Evaluation is based on retrieving 1000 documents. Across all 63 queries, there a total of 1253 relevant documents.

shows the number of queries where SRM exceeded bLM vs. the number of queries where performance was different. For example, 33/49 means that SRM out-performed bLM on 33 queries out of 63, underperformed on 49−33=16 queries, and had exactly the same performance on 63−49=14 queries. Bold figures indicate statistically significant differences (according to the sign test with $p < 0.05$).

The results show that SRM outperforms three baselines in the high-precision region, beating bLM's mean average precision by 29%. User-oriented metrics, such as R-precision and precision at 10 documents, are improved by 39.4% and 44.3% respectively. These improvements are statistically significant. The absolute performance figures are also very encouraging. Precision of 28% at rank 10 means that on average almost 3 out of the top 10 records in the ranked list are relevant, despite the fact that the requested fields are not available to the model.

We note that SRM continues to outperform bLM until the 100-document cutoff. After that, SRM degrades with respect to bLM. We feel the drop in effectiveness is of marginal interest because precision is already well below 10% and few users will be continuing to that depth in the list.

It is encouraging to see that SRM outperforms cLM, which is a cheating baseline that takes advantage of the field values that are supposed to be "missing". It is also encouraging to see that SRM outperforms bMatch, suggesting that best-match retrieval provides a superior strategy for selecting a set of appropriate training records.

5.8 Topic Detection and Tracking

All previously considered tasks were fairly intuitive – we did not need to spend much time motivating the ad-hoc retrieval process or the video searching scenario. Our final scenario is slightly different, and without some introduction it may appear somewhat artificial and confusing. In addition, topic detection does not really fit a strict definition of a retrieval scenario, where we typically have a user, her query and a static collection of documents. To provide a smooth transition from previous scenarios, we will spend a little bit of time introducing the big picture of the problem we are dealing with. A much more detailed exposition of Topic Detection and Tracking can be found in [1, 5, 104] and in the proceedings of the annual TDT workshop organized by NIST.

5.8.1 Definition

Topic Detection and Tracking is a research program dedicated to event-based organization of news reports. The idea is to take live news reports coming from newswire, television and radio sources, and organize them in a way that is intuitive and convenient to users. The focus of TDT is on dealing with live news, not with static collections. There are a number of distinct characteristics that make TDT very different from the tasks we previously described.

If we were to pick a single quality that makes TDT unusual, it would be the concept of *relevance* that is used by the program participants. Compared to other areas of information retrieval, TDT has an unusually clear and crisp definition of this fuzzy concept. The definition is based on a notion of *event*. An event is something that happens at a specific place and time and involves specific participants, be they human or otherwise. For example "the 1998 landing of hurricane Mitch" represents a specific event, whereas "deadly hurricanes" does not. A TDT *topic* embodies all activities directly related to a given *event*.

The term relevance is never used by the TDT program participants, but if it were used, it would mean that a particular story discusses the event in question, or discusses some activity directly related to that event. How does this relevance fit into the definitions of relevance discussed in chapter 2? By and large, it still fits into the bounds we delineated for this book. TDT relevance is concerned with topicality, rather than usefulness to some user task. TDT is non-interactive, any user feedback must happen before the TDT system starts operating. TDT deals with full-text documents, and as we will see these documents are quite varied in their nature. On the other hand, there are no queries in TDT, and this makes the problem quite interesting. In TDT, an information need is never expressed as a request, it can appear only in three possible forms: **(i)** it can be exemplified by a small set of on-topic stories (tracking task), **(ii)** it exists only as an unobserved variable, (clustering and link detection tasks), and **(iii)** it is *topical novelty*, i.e. the user is interested in things that have never been previously reported (new event detection task).

Another way in which TDT is dramatically different from other retrieval scenarios is the type of user for which the technology is targeted. A typical retrieval scenario, such as ad-hoc retrieval or video search, will target a casual user who will not expend a lot of effort formulating his request, and will likely be satisfied with a small number of relevant hits. In TDT the target user is quite different. He or she is a professional analyst interested in seeing *all* news reports related to a particular event. One could say that the target TDT user is much more recall-oriented than a typical web user. Because of this difference, TDT evaluation metrics are very different from the more traditional precision and recall.

TDT systems are supposed to operate on live news feeds, and that presents additional challenges of algorithmic design. In all previous scenarios we were dealing with a static collection of documents. The main advantage of a static collection is that we can estimate a statistical model once, and then use it for all queries issued against that collection. In TDT we do not have the same luxury. Whatever model we come up with has to grow and adapt to match the ever-changing nature of a live news-feed. Fortunately for us, adaptation and expansion is straightforward with the kernel-based models we proposed in the previous chapter. The on-line nature of TDT leads to another complication. In all previous scenarios it was sufficient to *rank* the documents in our collection, and the user himself would decide when it is time to stop examining the ranked list. In TDT we cannot afford the same process. The user needs to be alerted to events of interest as soon as possible, so for every document in the stream we need to make a hard decision of whether this document is relevant or not.

Last but not least, a TDT system has to deal successfully with very heterogeneous streams of data. Existing TDT collections represent a mix of newswire reports (e.g. Associated Press), editorial articles (e.g. New York Times), radio broadcasts (Voice of America) and TV news shows (CNN, ABC, NileTV). Newswire and editorials come in a nicely delineated textual form, but radio and TV shows come as a continuous stream of audio. TDT participants have access to a textual transcript of the audio, either as closed captions or as the output of the Automatic Speech Recognition system (ASR). The quality of transcribed text is substantially lower than the quality of newswires: for example names and locations are often garbled or spelled incorrectly. TDT participants have an option of working with the audio signal instead of the transcript. Another complication of TDT is its multi-lingual nature. The program started by focusing exclusively on English news, but over the years grew to include Chinese and Arabic sources. To make the task particularly challenging, TDT specifically focuses on events that receive news coverage in all three languages.

The TDT initiative defines five tasks, each of which represents one aspect of a complete news processing system. These tasks are:

1. **Story Segmentation.** Live news reporting comes in a form of a continuous stream of text. It is not broken down into cohesive units that we are used to working with. News anchors typically give very little indication that they are switching from one story to the next. This is not problematic for humans, who are naturally good at handling rapid semantic shifts, but it does pose a formidable challenge to automatic systems. The first task of TDT is to take a continuous stream of text and break it down into meaningful chunks, where each chunk discusses a particular topic, distinctly different from the preceding and the following chunks. These chunks are called *stories* and form the basic processing unit for the other four tasks.
2. **New Event Detection.** The purpose of this task (NED) is to identify the first reporting of some new event. For example, when an earthquake strikes, we want to flag the very first report of that event as *new*. Every subsequent report discussing the same earthquake would be marked as *old*. New Event Detection is a very interesting problem, in a way it directly addresses the problem of redundancy we discussed in section 2.1.5. The problem turns out to be very challenging: NED error rates remain high despite years of dedicated efforts by the researchers. A possible reason for that is that NED has no scope: it provides no intuition for what we *should* look for in a report; the only thing we know is what we *should not* look for: we should not retrieve anything we have seen before. Allan, Lavrenko and Jin [7] presented a formal argument showing that the New Event Detection problem cannot be solved using existing methods.
3. **Topic Tracking.** In the tracking task we start with a small set of stories discussing a particular event; the goal is to find all subsequent reports addressing the same event. Tracking is a step that naturally follows NED: once we have spotted some new interesting event, we want to track it in the news. The tracking task is the only TDT task where we are given explicit examples of the topic we are looking for. It is similar to information filtering or routing. The main differences are: (i) the number of training stories is very small: typically 1 to 4, (ii) we need to track a specific *event*, not the general *subject*, and (iii) we need to track in multiple languages and across multiple media.
4. **Topic Detection.** This task is also known as the *clustering* task, because the goal is to organize the news stream into a set of clusters, such that every cluster would contain stories discussing some particular event. What

makes the detection task very different from a normal clustering task is that we have to operate in a live, on-line environment. As soon as a story comes off the wire we have to either assign it to one of the existing clusters or create a new cluster if the story addresses a new, previously unseen event. The on-line nature of the task automatically rules out the use of a large number of popular clustering algorithms: such as top-down partitioning techniques, or agglomerative methods. Another way to look at the detection task is as a combination of the NED and the tracking tasks: first we detect a new event in the news stream, then we track the event using the first story as a single training example.

5. **Link Detection.** The last TDT task appears extremely simple and does not have the intuitive appeal of the previous tasks. We are given two stories and asked to decide whether or not they discuss the same event. We get no information regarding the type of event. The stories can come from different sources (e.g. newswire and radio), and can be written in two different languages. Despite its apparent simplicity, the link detection task is a very important part of TDT. We can view it as a component technology that is required by every other task. It is fairly obvious that a perfect solution to the link detection task would automatically solve all other TDT tasks. But we can make an even stronger claim: in our opinion, it is not possible to improve the current performance of NED, tracking and detection systems, unless we substantially improve the accuracy of link detection.[10]

5.8.2 Representation

In light of its fundamental role, we are going to focus all our TDT experiments on the Link Detection task. The task has the following setup. We are given two stories, $\mathbf{d}^1$ and $\mathbf{d}^2$; our goal is to decide whether the two stories discuss the same topic. The stories are selected randomly from the news stream; without loss of generality we will assume that $\mathbf{d}^2$ was released after $\mathbf{d}^1$. We have a collection of documents $\mathcal{C}$, which represents the news stream up to $\mathbf{d}^2$. For simplicity, we will assume that all documents in the collection are written in the same language. When they are not, we assume that original documents are mapped to some common language using a statistical dictionary (see section 5.3.1 for details).

The details will closely follow the relevance feedback scenario (section 5.2). The latent representation X of each story consists of a string length, followed by $M-1$ words from the common vocabulary $\mathcal{V}$. The document generating transform will operate exactly as in the ad-hoc case (equation 5.1): it will return the first X_1 words from X. There are no queries in TDT, so the query

[10] A possible exception to this claim is the tracking task with a large number of training stories. In that case it may be possible to improve performance in a way that cannot be traced back to improved link detection accuracy.

transform seems unnecessary. However, in order to avoid a certain problematic behavior of our model we will introduce something quite similar to a query transform; the details will be provided later. As before, we will take string length to be uniformly distributed on $1 \ldots M-1$; the parameter space Θ will be the word simplex $\mathbb{P}_{\mathcal{V}}$.

5.8.3 Link detection algorithm

Our approach to Link Detection will be as follows. Given the partial news stream $\mathcal{C}$ and the two stories $\mathbf{d}^1$ and $\mathbf{d}^2$ we will:

1. convert the stories $\mathbf{d}^1, \mathbf{d}^2$ into query-like representations $\mathbf{q}^1, \mathbf{q}^2$
2. use the news stream $\mathcal{C}$ to estimate relevance models RM_1 and RM_2
3. compute a modified form of cross-entropy between the resulting parameter vectors
4. if cross-entropy falls below a certain threshold, conclude that $\mathbf{d}^1$ and $\mathbf{d}^2$ discuss the same topic; if entropy is above the threshold – stories talk about different events

The overall process is quite similar to the ad-hoc and relevance feedback scenarios. Nevertheless, the nature of TDT leads to a few interesting changes.

Transforming documents into queries

The first change involves transforming the original story $\mathbf{d}^1$ into a query-like representation $\mathbf{q}^1$; the same is done for $\mathbf{d}^2$. We are going to define a query generator that attempts to pick out the most salient key-words from $\mathbf{d}^1$. There exist a large number of IR heuristics that can help us achieve this goal. For example, from the standpoint of the 2-Poisson indexing model [52], we simply want to find *elite* words for $\mathbf{d}^1$. We are going to approach the problem in a slightly different fashion and use the hypergeometric distribution instead of the Poisson. To tease out the salient words we will ask the following question for every word v:

Suppose we randomly draw n words from the entire stream $\mathcal{C}$.
What is the chance we will observe n_v instances of v?

Here n is the length of $\mathbf{d}^1$, and n_v is the number of times the word v was observed in $\mathbf{d}^1$. We can answer the above question with a *hypergeometric distribution*, which assumes sampling from $\mathcal{C}$ without replacement. Let N_v denote the total number of times v occurs in the news stream, and take N to be the total number of words in the stream. The likelihood of getting n_v instances of v by pure chance is:

$$P_{chance}(n(v, \mathbf{d}^1)) = \binom{N_v}{n_v} \binom{N - N_v}{n - n_v} / \binom{N}{n} \tag{5.38}$$

In order to get the most salient words, we pick 10 words that have the *lowest* $P_{chance}(n_v)$, i.e. the query $\mathbf{q}^1$ will be composed of the words that are *least* likely to occur in $\mathbf{d}^1$ under the background statistics. The same procedure is applied to $\mathbf{d}^2$ to get the query $\mathbf{q}^2$.

There are two motivations for taking the admittedly heuristic step of collapsing a story to a handful of keywords. The first is computational complexity: reducing the number of words will lead to a significantly faster computation of relevance models RM_1 and RM_2. The number of query words forms a bottleneck in the algorithmic optimizations we use to compute equation (5.6): estimating a posterior for a 1000-word document takes 100 times longer than a similar posterior for a 10-word query. Computational complexity is all the more important because in link detection we have to repeat the estimation tens of thousands of times – one for each story in each testing pair. Contrast that with previous scenarios, which involved at most 100-200 queries each.

The second motivation comes from the observations we made in the relevance feedback scenario (section 5.2). Recall that long observations $\mathbf{r}$ lead to a particularly simple expression for the relevance model $RM_{\mathbf{r}}$ – the form given by equation (5.15). In our case, multiplying together thousands of probabilities (one for every word in $\mathbf{d}^1$) would make the posterior density $p(\theta|\mathbf{d}^1)$ spiked over a single point $u_{\mathbf{d}^1}$. As a result, the relevance model RM_1 will be identical to the empirical distribution of words in $\mathbf{d}^1$. Informally, this means that our estimation efforts have no effect – we end up with the original document $\mathbf{d}^1$, with none of the *synonymy* or *query expansion* effects that gave us a boost in ad-hoc retrieval performance. This is clearly undesirable. Reducing the story $\mathbf{d}^1$ to a small set of keywords $\mathbf{q}^1$ allows us to steer around the problem. The posterior density $p(\theta|\mathbf{q}^1)$ will be spread out over more points in $\mathbb{P}_{\mathcal{V}}$, and the resulting estimate will be a mixture of many related stories. The story $\mathbf{d}^1$ itself will be a highly weighted element of this mixture, the important point is that it will not be the *only* element.

Modified relative entropy

After reducing each story to its keyword representation we can compute the relevance models RM_1 and RM_2 the same way as in the ad-hoc scenario: using equation (5.6). The next step is to measure the dissimilarity of the two vectors. Cross-entropy is a natural measure of dissimilarity for distributions, and it served us well in the previous scenarios. However, link detection is different from earlier tasks in one very important respect: it is a *symmetric* problem. In all previous cases we had a single query $\mathbf{q}$ and had to rank documents with respect to that query. In a way, the query served as a fixed point for the scoring procedure: one side in the cross-entropy was always fixed. In link detection we have no query to act as a fixed point: both sides are allowed to vary. Our dissimilarity measure should be symmetric to reflect this fact. After much experimentation we settled on the following measure:

$$D_{lnk}(\mathbf{d}^1, \mathbf{d}^2) = \sum_{v \in \mathcal{V}} RM_1(v) \log \frac{BG(v)}{RM_2(v)} + \sum_{v \in \mathcal{V}} RM_2(v) \log \frac{BG(v)}{RM_1(v)} \quad (5.39)$$

An intuition for equation (5.39) is as follows. We start with the relative entropy $KL(RM_1||RM_2)$, which gives the number of bits wasted in "encoding" story 1 using story 2. We normalize it by subtracting $KL(RM_1||BG)$ – the number of bits wasted by "encoding" story 1 with the background distribution. The quantity $KL(RM_1||BG)$ is meant to compensate for the overall difficulty of encoding story 1. In Information Retrieval the quantity is known as the *clarity* measure[37]; it is used to gauge the degree of ambiguity inherent in the user's query. After subtracting the clarity we get $KL(RM_1||RM_2) - KL(RM_1||BG)$, which is the number of bits wasted by encoding story 1 with story 2 as opposed to the background distribution. Now we can repeat the argument in the opposite direction, encoding story 2 with story 1, and comparing to the background. The final step is to add the relative number of bits wasted in encoding story 1 and then story 2 – the left and right summations of equation (5.39). We can also relate the dissimilarity measure given by equation (5.39) to the probability ratio discussed in section 3.6.1. The left summation bears similarity to the probability ratio for observing story $\mathbf{d}^1$ under the relevance model of story $\mathbf{d}^2$; the only difference is that probabilities $RM_1(v)$ are used in place of actual frequencies $n(v, \mathbf{d})$. The right summation has a similar interpretation.

The last part of the link detection process is to compare the dissimilarity measure $D_{lnk}(\mathbf{d}^1, \mathbf{d}^2)$ to a threshold value. If dissimilarity exceeds the threshold we conclude that the stories discuss different events. If dissimilarity is below the threshold, we decide that $\mathbf{d}^1$ and $\mathbf{d}^2$ talk about the same event.

5.8.4 Experiments

In this section we evaluate performance of relevance models, as described above, on the Link Detection task of TDT. First, we describe the experimental setup and the evaluation methodology. Then we provide empirical support for using a modified form of relative entropy. Finally we show that relevance models significantly outperform simple language models, as well as other heuristic techniques. The experiments discussed in this section were previously reported in [2, 74].

Datasets and Topics

The experiments in this section were performed on three different datasets: TDT2, TDT3 and TDT4 [24]. Most of the training was done on a 4-month subset of the TDT2 dataset. The corpus contains 40,000 news stories totaling around 10 million words. The news stories were collected from six different sources: two newswire sources (Associated Press and New York Times), two radio sources (Voice of America and Public Radio International), and two television sources (CNN and ABC). The stories cover January through April of 1998. Radio and television sources were manually transcribed at closed-caption quality. Testing was carried out on the TDT3 and TDT4 datasets, containing 67,111 and 98,245 news stories respectively. The total number of words was around 21 million for TDT3 and 39 million for TDT4. Both datasets contain English, Arabic and Chinese stories as newswire, radio and television reports. In a pre-processing stage, all English stories were stemmed using a dictionary-based stemmer, and 400 stop-words from the InQuery [3] stop-list were removed.

Human annotators identified a total of 96 topics in the TDT2 dataset, ranging from the 1998 Asian financial crisis, to the Monica Lewinsky scandal, to an execution of Karla Faye Tucker. Each topic is centered around a specific event, which occurs in a specific place and time, with specific people involved. 56 out of these 96 topics are sufficiently represented in the first four months of 1998 and will form the basis for our evaluation. Similar annotations were done on the TDT3 and TDT4 datasets.

Evaluation Paradigm

The system is evaluated in terms of its ability to detect the pairs of stories that discuss the same topic. A total of 6363 story pairs were drawn from the dataset (according to the official TDT2 sampling). 1469 of these were manually [24] judged to be on-target (discussing the same topic), and 4894 were judged off-target (discussing different topics). During evaluation the Link Detection System emits a YES or NO decision for each story pair. If our system emits a YES for an off-target pair, we get a *False Alarm* error; if the system emits a NO for on-target pair, we get a *Miss* error. Otherwise the system is correct. Link Detection is evaluated in terms of the detection cost [45], which is a weighted sum of probabilities of getting a Miss and False Alarm:

$$Cost = P(Miss) \cdot C_{Miss} + P(FA) \cdot C_{FA}$$

In current evaluations of Link Detection, C_{Miss} is typically set to 10×0.02, and $C_{FA} = 1 \times 0.98$. Note that by always answering YES a system would have no misses and therefore a cost of 0.2 (similarly, always answering NO guarantees a cost of 0.98). To penalize systems for doing no better than a simple strategy like that, the cost is normalized by dividing by the minimum of those two values (here, 0.2). A normalized cost value near or above 1.0 reflects a system that performs no better than random. An operational Link Detection System requires a threshold selection strategy for making YES / NO decisions. However, in a research setting it has been a common practice to ignore on-line threshold selection and perform evaluations at the threshold that gives the best possible cost. All of our experiments report the minimum normalized detection cost: $Cost_{min}$. Before we proceed we would like to repeat that TDT is evaluated using a *cost* metric, not an accuracy metric. In all TDT experiments **lower means better**.

Evaluation of various entropy formulations

In Section 5.8.3 we described a modified relative entropy measure, and provided an argument for why we believe this measure may perform better than regular relative entropy. To evaluate the value of our modification we perform a simple experiment without constructing relevance models. Given a pair of stories A and B we construct empirical language models of each story, smooth them with the background model BG, and measure divergence. We consider four different divergence measures:

1. relative entropy (KL divergence): $KL_1(A, B) = KL(A||B)$
2. symmetric version of divergence: $KL_2(A, B) = KL(A||B) + KL(B||A)$
3. adjusted relative entropy: $KL_{c,1}(A, B) = KL(A||B) - KL(A||BG)$
4. symmetric adjusted entropy: $KL_{c,2}(A, B) = KL_{c,1}(A, B) + KL_{c,1}(B, A)$

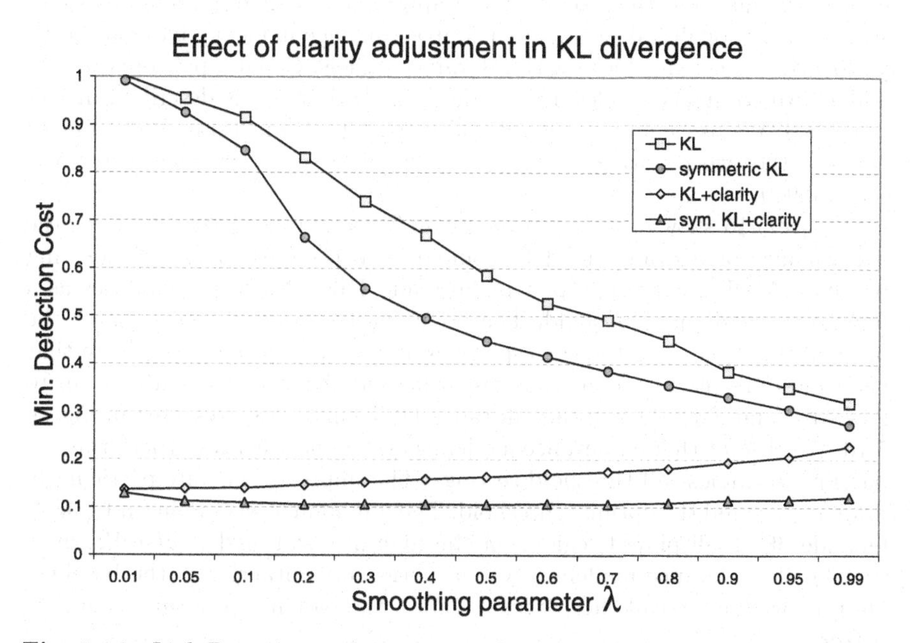

Fig. 5.14. Link Detection performance on the TDT2 dataset. Clarity-adjustment leads to significantly lower error rates. Symmetric versions of KL perform better than asymmetric versions. Symmetric KL with clarity adjustment is best and most stable with respect to the smoothing parameter λ.

Figure 5.14 shows the minimum detection cost ($Cost_{min}$) of the four measures as a function of the smoothing parameter λ from equation (4.29). We observe that clarity-adjusted KL leads to significantly lower errors for all values of λ. Clarity adjustment also leads to smaller dependency on λ, which makes tuning easier. We also note that for both simple and clarity-adjusted

KL, we get significantly better performance by using symmetric divergence. The best performance $Cost_{min} = 0.1057$ is achieved by using the symmetric version of clarity-adjusted KL when $\lambda = 0.2$. This performance will be used as a baseline in later comparisons with relevance models. The baseline is competitive with the state-of-the-art results reported in the official TDT evaluations [103].

Performance on the TDT2 / TDT3 datasets

Experiments discussed in this section were performed for the 2001 TDT evaluation and were originally reported in [74]. That evaluation used a slightly different technique for converting stories into queries. Instead of using the hypergeometric distribution to pick 10 *most unusual* words, in [74] we simply picked 30 *most frequent* words from each document. Overall detection cost was similar, but the procedure was approximately five times more expensive computationally. We compare a simple language modeling baseline to the performance we can achieve with relevance models. Given a pair of stories A and B, we construct a relevance model from each story as described in section 5.8.3. We use symmetric clarity-adjusted KL as a measure of divergence between the two resulting relevance models. The smoothing parameter λ is set to 0.999.

Figure 5.15 shows the *Detection Error Tradeoff* (DET) [86] curve for the performance of relevance models, compared to the best language modeling baseline. A DET curve is a plot of Miss and False Alarm probabilities as a function of a sliding threshold. The point marked on each curve marks the optimal threshold, and the corresponding minimum cost $Cost_{min}$. Note that the values are plotted on a Gaussian scale and that the axes only go up to 20% Miss and False Alarm; the full-range DET curves are presented in Figure 5.16. We observe that a relevance modeling system noticeably outperforms the baseline for almost all threshold settings. The improvements are particularly dramatic around the optimal threshold. The minimum cost is reduced by 33%. Outside of the displayed region, on the high-precision end ($FalseAlarm < 0.01\%$), the relevance modeling system noticeably outperforms the baseline. On the very high-recall end ($Miss < 1.5\%$), the baseline performs somewhat better.

The results in Figure 5.15 were achieved by careful tuning of parameters on the 4-month subset of the TDT-2 corpus and do not represent a blind evaluation. However, the same parameters were used in the official TDT 2001 blind evaluation on the 3-month TDT-3 corpus. The results in that case were comparable to those described above (see Figure 5.17 for details). The system based on Relevance Models significantly outperformed a state-of-the-art vector-space system (cosine with Okapi *tf.idf* weighting). The normalized minimum cost was reduced from 0.27 to 0.24. This suggests that our parameter settings generalized reasonably well to the new dataset and the new set of topics.

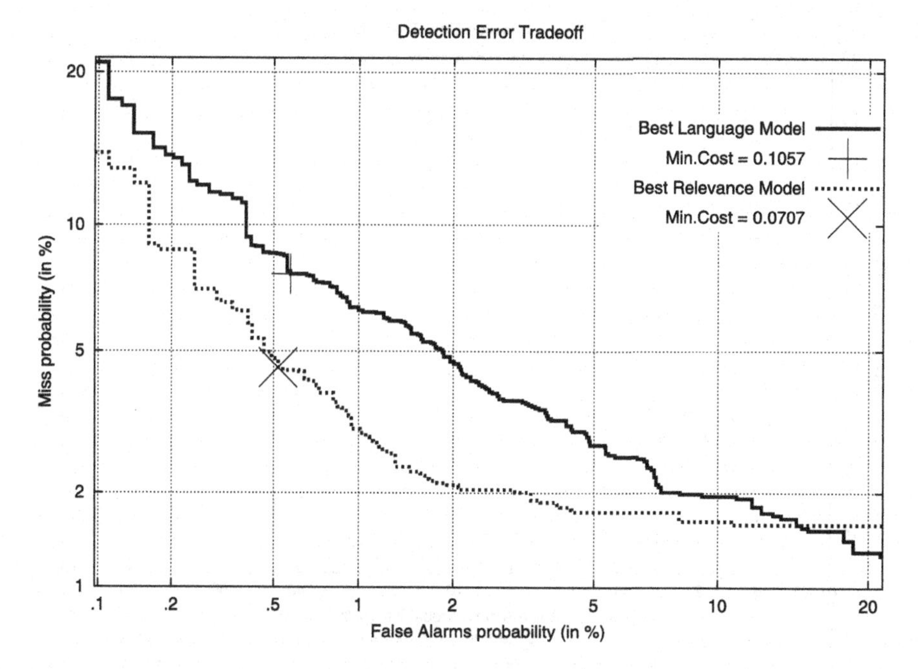

Fig. 5.15. Relevance model noticeably outperforms the baseline for all threshold settings in the region of interest. Minimum Detection Cost is reduced by 33%

Performance on the TDT3 / TDT4 datasets

We now turn our attention to performance of our model in the official TDT-2003 evaluation. Figure 5.18 shows the performance of the relevance model compared to the performance of our baseline system: vector-space, using cosine similarity with tf·idf weighting. Experiments were done on the 2003 training corpus (TDT3). Each pair of documents was compared in its own language when possible, all cross-language pairs were compared in machine-translated English. We observe that relevance model noticeably outperforms the vector-space baseline at all Miss levels. Table 5.17 provides a more detailed comparison of performance on both TDT-3 and TDT-4 datasets (pre-adjudicated results). We show results of using both a 1 database (where every pair is compared using machine-translated English), and 4 databases (where monolingual pairs are compared in their own language). We report the minimum cost for TDT-3 and both minimum and actual cost for TDT-4 (where we had to guess the optimal threshold).

We can draw several conclusions from the results in Table 5.17. First, we observe that performing comparisons in the native language (4db) is always

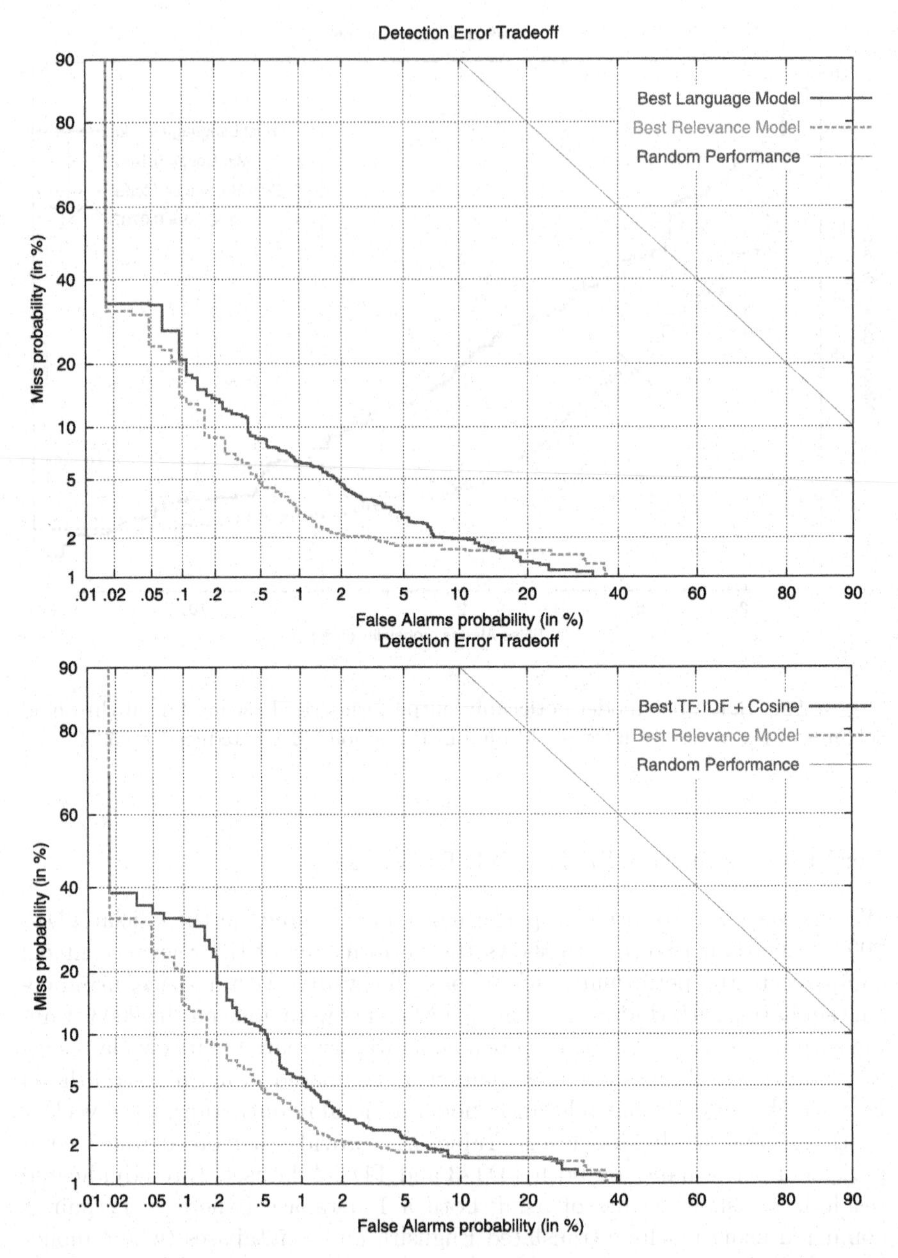

Fig. 5.16. Performance of relevance model on the 2001 training data (TDT2). Relevance model with optimal parameters outperforms both the optimal Language Modeling system (top), and the optimal vector-space system using Cosine with Okapi term weighting (bottom). Minimum Detection Cost was reduced by 33% and 25% respectively (not shown).

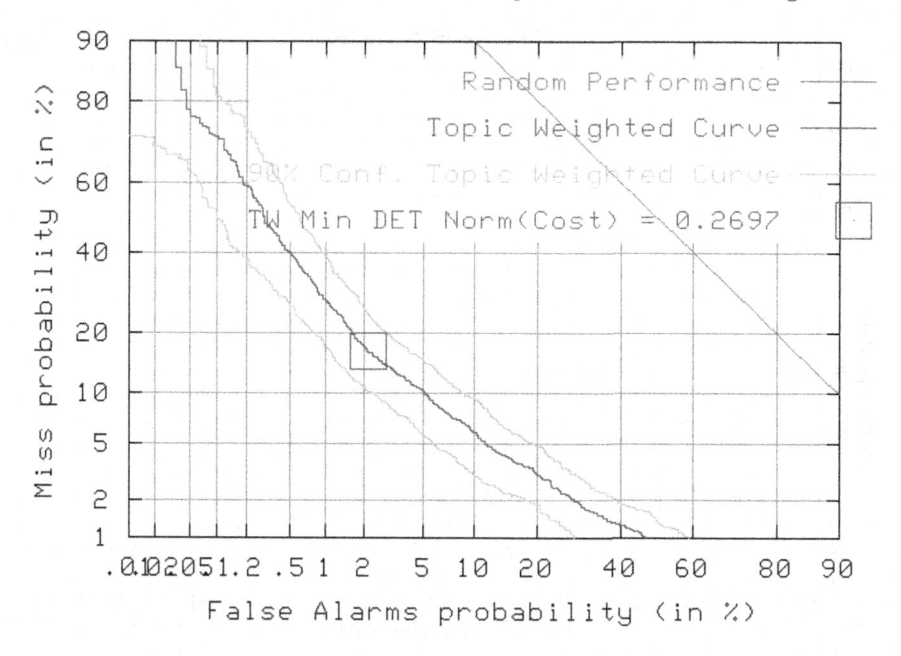

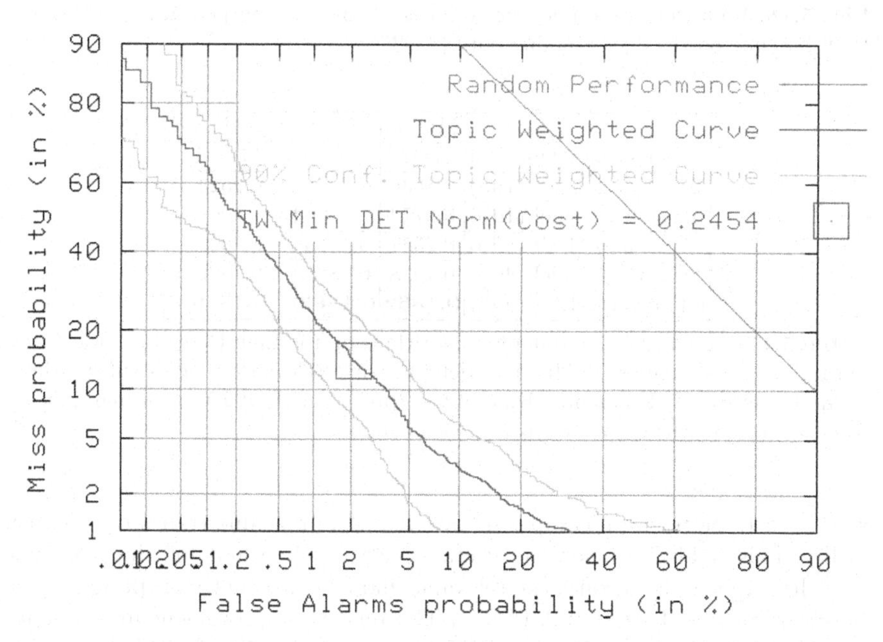

Fig. 5.17. Performance of relevance models in the official TDT 2001 evaluation (TDT3). All the parameters were tuned on the training dataset, and no part of the evaluation dataset was used prior to evaluation. Relevance model (bottom) consistently outperforms the vector-space model (top). Minimum Detection Cost is reduced by 10%.

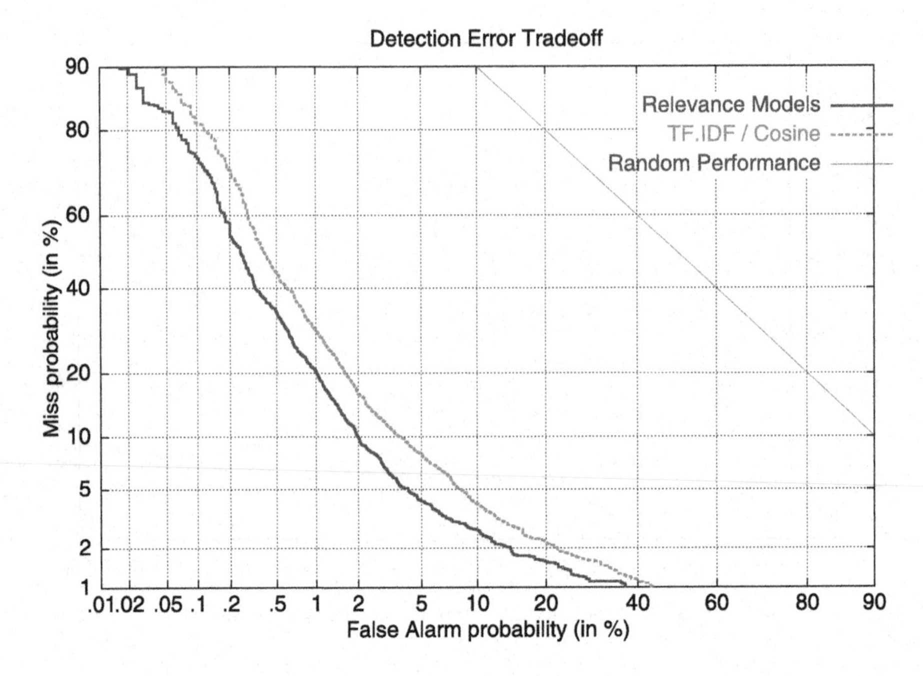

Fig. 5.18. Link Detection: relevance models consistently outperforms the cosine / tf·idf baseline on the 2003 training data (TDT3).

	Minimum Cost		Actual Cost
Detection Algorithm	TDT-3	TDT-4	TDT-4
Cosine / tf·idf (1db)	0.3536	0.2472	0.2523
relevance model (1db)	0.2774	—	—
Cosine / tf·idf (4db)	0.2551	0.1983	0.2000
relevance model (4db)	0.1938	0.1881	0.1892
rel.model + ROI (4db)	**0.1862**	**0.1863**	**0.1866**

Table 5.17. Performance of different algorithms on the Link Detection task. Comparing in native language (4db) is a definite win. Relevance models substantially outperforms the tf·idf baseline. Rule of Interpretation (ROI[2]) has a marginal effect. Boldface marks minimum cost in each column.

significantly better than comparing everything in machine-translated English (1db). This holds both for the vector-space baseline and for the relevance modeling approach. Second, the relevance model consistently outperforms the baseline both on TDT-3 and TDT-4. The relative improvement in minimum cost is a very substantial 27% on TDT-3 data, but only 7% on TDT-4, suggesting that smoothing parameters of the model were over-tuned to the training data. Our official submission to the TDT 2003 evaluation is represented by the last line in the table. *ROI* [2] refers to an attempt to further improve performance by adjusting story similarities depending on which *rule of interpretation* [24] they fall into. The adjustment has a very small but positive effect on performance. We observe that our threshold selection strategy for TDT-4 was quite successful: actual cost is quite close to the minimum cost. Bold figure mark the best run in each column. It is worth noting that in addition to improved performance, the present formulation of the relevance model for Link Detection is almost 80% faster than the formulation described in [74]. The computational speedup is mainly due to using a small subset of words as a "query".

Cross-modal evaluation

An essential part of TDT is being able to deal with multiple sources of news. The TDT2 corpus that was used in our experiments includes news from six different sources. Two of these (Associated Press and New York Times) are printed sources, the other represent broadcast news, which are transcribed from audio signal. Spoken text has very different properties compared to written sources, and part of the TDT challenge is development of the algorithms that can cope with source-related differences in reporting. To determine how well our algorithms perform on different source conditions, we partitioned the set of 6363 pairs into three subsets:

1. 2417 pairs where both stories come from a broadcast source; this set will be labeled "BN" (broadcast news)
2. 1027 pairs where both stories come from a printed source; this set will be labeled "NW" (newswire)
3. 2919 pairs where one story is from a broadcast source and the other from the printed source; this set will be labeled "NWxBN"

Figure 5.19 shows performance of the baseline and the relevance modeling systems on the three subsets we described. Performance is shown as a function of the smoothing parameter λ. First we observe that performance varies very significantly from one subset to another. Interestingly, both systems perform best on the "NWxBN" condition, even though it intuitively appears to be more challenging as we are dealing with two different language styles. Another very interesting issue is the value of λ that gives the best performance. Note that for the baseline system the optimal λ value is different for every condition: "BN" is optimized near $\lambda = 0.5$, "NW" - near $\lambda = 0.05$, while "NWxBN" is optimal near $\lambda = 0.7$. This means that for the baseline system we cannot select a single value of λ which will work well for all sources. In contrast to that, for the relevance modeling system all conditions are optimized if we set λ to 0.99, or any value close to 1. This is a very encouraging result, as it shows that relevance models are not very sensitive to source conditions.

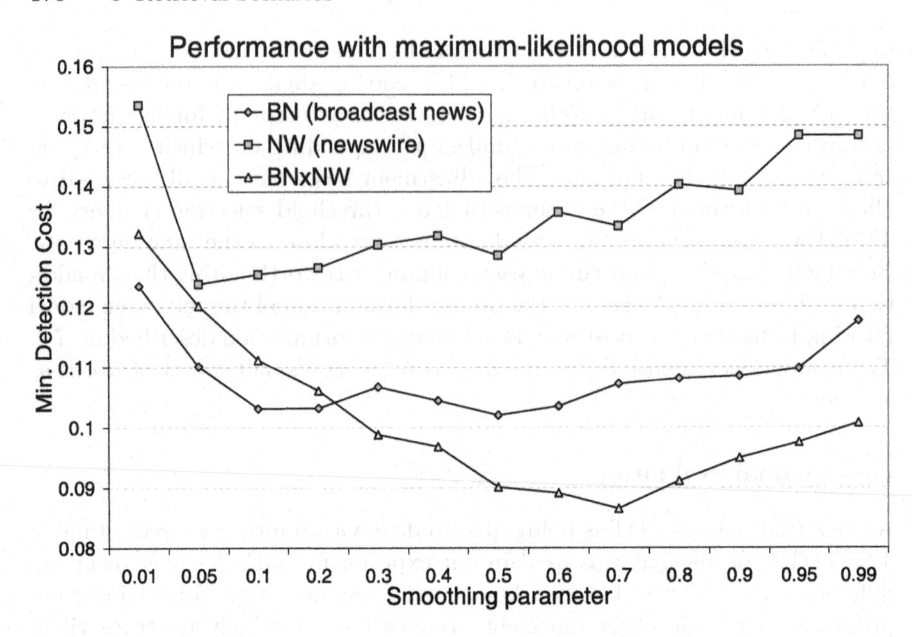

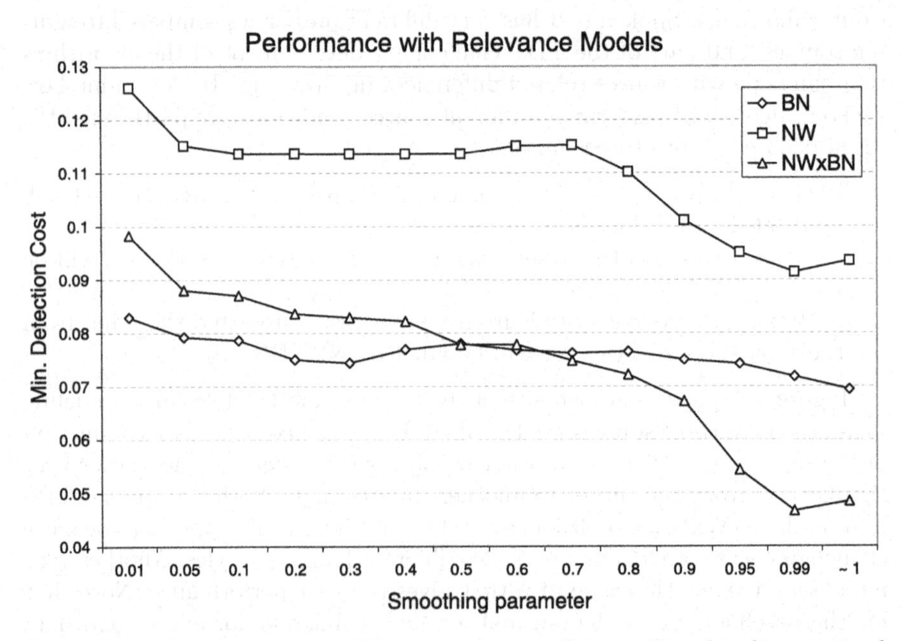

Fig. 5.19. Performance on different source conditions. Top: baseline, optimal smoothing value is different for every condition. Bottom: relevance models, all conditions are optimized as λ approaches 1.

6

Conclusion

Our sincere hope is that a reader will walk away from this book with a new way of thinking about relevance. We started this work by asserting that relevance serves as the cornerstone of Information Retrieval; we should perhaps have also said that relevance is the greatest stumbling block for the field. The fundamental task of information retrieval is to *retrieve relevant items in response to the query.* The main challenge lies in the fact that relevance is never observed. We know the query, we have the information items, but in most retrieval scenarios we will never be able to directly observe relevance, at least not until *after* we have already retrieved the items. In some sense this makes the fundamental retrieval task ill-defined. After all, how can we ever learn the concept of relevance if we have no examples to learn from? The answer is: we cannot, not unless we make some assumptions. These assumptions typically fall into two categories. In the first category we follow the classical probabilistic model [117] and come up with heuristics to approximate relevance. In the second category, e.g. language modeling [106], we get rid of relevance and develop a formal model of retrieval based on some hypothesized process that generates queries from documents, or vice versa. The goal of this book was to provide a third alternative.

We introduced a model of retrieval that treats relevance as a common generative process underlying *both* documents and queries. The centerpiece of our model is the generative relevance hypothesis, which states that for a given information need, relevant documents and relevant queries can be viewed as random samples from the same underlying model. To our knowledge, none of the existing retrieval models are based on a similar assumption. The generative hypothesis allows us to formally and naturally tie together relevance, documents and user queries. By treating both documents and queries as observable samples from the unknown relevance model, we are able to estimate all relevant probabilities without resorting to heuristics. In addition, assuming a common generative model allowed us to provide effective solutions for a large number of diverse retrieval problems. The generative approach allows us to associate words with pictures, video frames, and other words in different languages.

If theoretical arguments are not convincing, we would like to point out that the proposed model is very effective from an empirical standpoint. We have demonstrated that our generative model meets or exceeds state-of-the-art performance in the following retrieval scenarios:

1. **Ad-hoc retrieval.** Our model provides a formal approach to ad-hoc retrieval. We compared our generative model to four strong baselines: InQuery tf·idf [3], tf·idf with LCA [151], language models [106] and language models with query expansion [105]. Comparisons were carried out on five different datasets: WSJ, AP, FT, LA, TDT2. The datasets contain a total of over 800,000 documents and 450 queries. Our model outperforms the standard tf·idf baseline by 15-25% on all datasets. Most of the improvements are statistically significant. The model exceeds performance of the other three baselines by 5-10%.
2. **Relevance feedback.** The process of relevance feedback is very natural in our model: relevant examples are treated as samples from the underlying relevance model – in the same way as the query. We compared performance of our generative model to language models with relevance feedback [105]. Experiments were repeated on three different datasets (AP,FT,LA) and involved a total of 250 queries. The generative model performed slightly better than the baseline, although the improvement was not statistically significant.
3. **Cross-language retrieval.** The proposed generative model can easily learn to associate words in different languages. We described an effective way to retrieve Chinese documents in response to English queries. We tested our model on the TREC-9 dataset with 25 queries. We were able to outperform the state-of-the-art translation model [154] by 10% on long and short queries. Our model achieved 96-98% of the strong mono-lingual performance.
4. **Handwriting retrieval.** We demonstrated that our model can associate English words with manuscripts that are not susceptible to OCR. The model engendered the first operational system for searching manuscript collections with text queries. Our model achieves over 80% precision at rank 1 on a collection of George Washington's handwritings. Mean average precision is 54%, 63%, 78% and 89% for 1-, 2-, 3- and 4-word queries respectively. There are no baselines we could compare to.
5. **Image retrieval.** A continuous-space extension of our model represents the current best-performing system for automatic annotation of images with keywords. We tested the model on the Corel benchmark dataset [41] and found it to outperform the best published results by up to 60%. The model provides a very effective means for retrieving unlabeled images in response to multi-word textual queries. Mean average precision is 23%, 25%, 31% and 44% for 1-, 2-, 3- and 4-word queries respectively.

6. **Video retrieval.** The same continuous-space model facilitates text-based searching through larger collections of video footage. In our experiments the model achieves mean average precision of 30%, 26%, and 32% for 1-, 2-, and 3-word queries respectively.
7. **Structured search with missing data.** We have demonstrated how the generative model can be applied to the problem of searching structured databases with missing field values. Relevance models constructed from the user's query allow us to infer plausible values for missing fields, which is particularly valuable in semi-structured collections. We observe 29% improvement in mean average precision over several realistic baseline strategies.
8. **Topic Detection and Tracking.** Last, but not least, the generative model results in a very significant improvement in the accuracy of link detection – the core technology behind TDT. Under in-house evaluations, our model yields a very substantial 25-30% improvement over the best baseline systems. More importantly, our model outperforms the same baselines by 5-7% in completely blind annual evaluations carried out by NIST. Our model was the best performing link detection system in the 2003 TDT evaluation.

The second contribution of our work is a new and effective generative process for modeling discrete exchangeable sequences (bags of features). In our opinion, this process may be of interested to a much wider audience, beyond the broad field of Information Retrieval. Nothing in the model is specific to language, to documents and queries – it can be applied whenever one is trying to learn something about unordered discrete observations, e.g. feature sets. Our formulation was motivated in no small part by the work of Blei, Ng and Jordan [12]. We view their work as a very important development in discrete generative models, but we strongly disagree with the structural assumptions made in their LDA model, the same structural assumptions that previously led Hoffman to develop the pLSI model.

Specifically, we believe it is a bad idea to represent text collections in terms of a small number of factors. Natural language is an incredibly complex phenomenon, and in our experience it is best to think of it as a myriad of outliers, rather than a handful of clusters with some Gaussian or Dirichlet noise. Any type of dimensionality reduction, be it clustering, information bottleneck methods, principal component analysis, mixture models, pLSI or LDA will invariably destroy rare events, which are so important for retrieval applications.

As we mentioned before, we do not argue with the notion that the *dimensionality* of text is lower than the number of words in the vocabulary. We believe it is quite plausible that documents in a given collection really do form a low-dimensional sub-space. However, we strongly disagree with the conception that this sub-space may be linear, or even convex. It is our personal view that collections of documents are stringy, fibrous, contagious in nature: an author writes a story, then someone picks it up and writes a response, the author revises the story, then someone discovers new facts relevant to it. The original story morphs and evolves in a step-by-step fashion, each step being taken not too far the predecessor, but after a number of steps we may end up quite far from where we started. We believe it highly unlikely that the majority of these paths will revolve around a certain point in space – the point which we as humans would like to call the *topic* of discussion.

We wanted to create a model that would be able to capture these paths in their original form, rather than cramming them into a low-dimensional polytope defined by cluster centroids, principal components or pLSI aspects. To that end we proposed a kernel-based density allocation model, which can be thought of as a series of spheres, placed along the paths along which the stories develop. In our opinion, the model is simple and effective. It is very easy to estimate, and naturally allows for discriminative tuning when the goal is to achieve a good performance on a specific task. The most important quality of the model is that it makes no structural assumptions about the data. Every training example is preserved and has a say when we predict the likelihood of new events. We have devoted chapter 5 to showing that this kernel-based approach is very effective in seven very different retrieval scenarios.

Finally, in the course of writing this book we uncovered a number of very curious new facts about existing models. These facts are:

- contrary to popular belief, the classical probabilistic model is not based on the assumption of word independence. Instead, it requires a much more plausible assumption of *proportional interdependence* (section 2.3.2).
- there is a good reason why explicit models of word dependence have never resulted in consistent improvements in retrieval performance, either in the classical or the language-modeling framework (sections 2.3.2 and 2.3.3).
- documents and queries can be viewed as samples from the same underlying distribution, even though they may look very different (sections 3.4.2 and 3.4.3).
- probability ratio, is not the best criterion for ranking documents; relative entropy is a better alternative (section 3.6.3).
- a ganerative probabilistic LSI model with k aspects is equivalent to a regular, document-level mixture model with $N >> k$ components; the original pLSI is equivalent to an "oracle" unigram model (section 4.3.4).
- Latent Dirichlet Allocation is not a word-level mixture, it is a regular document-level mixture. One can think of it as a simple Dirichlet model restricted to the k-topic sub-simplex (section 4.3.5).
- one should be very cautious when analyzing graphical models, as they can be quite misleading (section 4.3.6).

6.1 Limitations of our Work

We can gain further understanding of our model by looking at the cases that cannot be handled with our methods. We will highlight three limitations that we consider to be the most important: **(i)** the closed-world nature of our retrieval model, **(ii)** the assumption of exchangeability, and **(iii)** the computational cost of our model.

6.1.1 Closed-universe approach

Our model of Information Retrieval represents a closed-universe approach. We assume that there exists a single information need R, along with an associated relevant population $\mathcal{S}_R$ in the representation space. According to the generative hypothesis, each relevant document is a sample drawn from $\mathcal{S}_R$, while each non-relevant document is a sample from some other population. This works well as long as we have only one information need. But suppose we are dealing with two distinct information needs R_1 and R_2, and some document $\mathbf{d}$ happens to be relevant to both of them. In this case we would run into a philosophical inconsistency: according to the GRH, $\mathbf{d}$ must have been drawn from $\mathcal{S}_{R_1}$, since it is relevant to R_1. But $\mathbf{d}$ is also relevant to R_2, so it must be an element of $\mathcal{S}_{R_2}$, which implies it cannot be a sample from $\mathcal{S}_{R_1}$, unless the two populations are allowed to overlap. We would like to point out that the same kind of inconsistency is present in the classical probabilistic model [117]. Furthermore, Robertson and Hiemstra [111] argue that an even greater inconsistency lies at the foundation of the language-modeling framework: assuming that the query $\mathbf{q}$ is a sample drawn from a relevant document implies that there may be only one relevant document in the entire collection, namely the document $\mathbf{d}^*$ that was the source of $\mathbf{q}$.

The important thing to realize that the inconsistencies described above are purely philosophical. They have absolutely no practical impact since we are dealing with *probabilities* that a given sample ($\mathbf{d}$ or $\mathbf{q}$) would be drawn from a given language model, and not with hard yes-or-no decisions about the origin of the sample. If we insist on a philosophically consistent model, the solution is to define a new universe (event space) for every new information need. The same closed-universe approach is characteristic of all existing probabilistic retrieval models, with a possible exception of the *unified* model [14, 109].

6.1.2 Exchangeable data

Our generative model is based on the assumption that representation components (words) $X_1 \dots X_n$ are exchangeable. As we discussed in section 3.4, the assumption of exchangeability is much weaker than the popular assumption of word independence. However, it is an assumption about the data, and thus constitutes a limitation of our model. Our model can only be applied in the cases where we can reasonably assume that the order of features (words) is irrelevant. In chapter 5 we demonstrated that the assuming exchangeability works for a surprisingly wide array of possible tasks, but we can certainly imagine scenarios where the model would not perform well.

A good example of a domain where our model might fail is *music retrieval.* Suppose we have a collection of musical pieces, where each piece is expressed as a sequence of notes. The semantics of each musical piece is embodied not in the individual notes, but rather in their *sequence.* Assuming exchangeability of notes in a piece would prove disastrous: individual notes are completely meaningless when taken out of context. It would be equivalent to assuming that individual *characters* in text are exchangeable.

Another domain where our model may not fare well is hierarchical data. Suppose we want to perform retrieval tasks on a *treebank* corpus, where individual documents represent parse trees. Leaves of a given parse tree are not as sequence-dependent as notes in a musical piece: in many grammars we can re-order the sub-trees and still retain the meaning. However, we would lose a lot of information if we assumed the leaves to be completely exchangeable.

Fortunately, in many practical scenarios we can construct an exchangeable representation that can approximate order-dependent data to some degree. In music retrieval we may consider using n-gram sequences rather than individual notes; exchangeability of these n-grams is a more realistic assumption. Similarly, in the case of hierarchical data, we could consider the entire path from a root to the leaf as a single "word", and then assume that these paths are exchangeable.

6.1.3 Computational complexity

A serious limitation of our model is the relatively high computational expense associated with kernel-based estimates. As we discussed in section 4.4.3, the number of parameters in our model grows linearly with the size of the training set. These are not free parameters, so overfitting is not a grave concern, but the sheer number of them makes the model inefficient relative to the baselines. For certain scenarios the added computational expense is not justified by a relatively small improvement in performance. For example, we report 20-25% improvement in average precision for the ad-hoc retrieval scenario. While these improvements are substantial and statistically significant, they do not represent a groundbreaking leap in retrieval accuracy: we may gain one or two relevant documents in the top 10 results. At the same time, the computational expense of our generative model is an order of magnitude higher than the expense of a simple word-matching baseline. In this case a 25% improvement in precision is probably not worth the added cost.

In other retrieval cases, the situation is quite different. For example, in the image retrieval scenario there are no simple and effective baselines. Any operational system has to be able to associate bitmap images with textual captions, which is not a trivial process. State-of-the-art baselines are based on statistical translation and maximum-entropy models. The combined (training

+ testing) complexity of these models is comparable to the overall expense of using our generative model. And even if the computational cost was higher, the 60% improvement in accuracy would certainly justify the added expense.

6.2 Directions for Future Research

Our work is far from complete. The ideas presented in this book can be extended in a number of important ways, and applied in many scenarios beyond the ones we described. In the remainder of this chapter we will briefly outline the directions that we consider most promising, the directions that we hope to explore in our future work.

6.2.1 Relevance-based indexing

We view computational complexity as the single greatest obstacle to a wide adoption of the model proposed in this book. As a kernel-based approach, relevance models require $O(n)$ computations at the time when retrieval or annotation needs to be performed, n being the size of the training dataset. This complexity becomes prohibitive when dealing with web-sized collections. Further complicating the issue, relevance models are not easily amenable to standard IR indexing methods, which rely on a sparse nature of language: the vast majority of indexing features (words) are absent from any given document. Relevance models are by definition non-sparse, and no simple heuristic allows us to decompose it into a *signal+noise* form, which would make it amenable to traditional indexing methods. Fortunately, there are two directions which appear promising in addressing the issue:

1. **Localized indexing.** We are experimenting with a technique for augmenting the traditional feature-based indices with nodes adjacent to the document in the locality graph. Preliminary results suggest that the new indexing structure enables relevance models to operate with the same speed as the baseline retrieval methods.
2. **Approximate matching.** We are investigating multi-pass algorithms for computing the similarities between a relevance model and the documents in the collection. The basic idea is to rapidly reduce the collection to a small set of possibly-similar documents, and then re-score that set.

6.2.2 Hyper-linked and relational data

We believe that our approach has natural extensions to hyper-linked and relational environments. Consider a very simple case of relational structure: *hyperlinks*. Every document d in our collection may have outgoing links to some set of documents $d' \in \eta_d$. The set η_d may be called the set of *neighbors* of d. Is there a way we can model this sort of structure within our generative model? We believe that we can, and the extension is surprisingly straightforward. Let $L_1 \ldots L_k$ denote the k links going out of a document. Each L_i is a multinomial random variable, its possible values are the identifiers of documents in our collection. For example, the event L_1=23 would mean that the first out-going link points to the twenty-third document in the collection. The set of all possible identifiers forms a finite vocabulary $\mathcal{L}$. At a conceptual level, this vocabulary is no different from a vocabulary of a real language. We can pretend that $L_1 \ldots L_k$ represent words in some foreign language and use the same approach that was described in section 5.3. Given a set of hyper-linked documents we can learn the associations between the English words and the outgoing links. Then, given a new un-linked document, we can attempt to predict which existing documents it might point to. Note that the same procedure can be used to learn associations between the content of a document and links pointing *into* it. In other words, we should be able to predict which documents are likely to point into the new item. Equivalently, given a set of incoming links, we will be able to predict a set of English words that are likely to occur in the page.

Relational structures

Once we are able to model the simple link structure, the extension to relational data is very natural. The distinguishing feature of relational databases is the fact that each link has a *type* associated with it. For example, consider a university personnel database. Each entity represents a person, and these entities can be linked in a number of different ways. We may have *course_instructor* links going from a professor to a large group of students, as well as *academic_advisor* links going to a much smaller group. We can also have *shared_office* links between the students and *shared_grant* links between the professors. We can model relational links associated with each person by a set of random variables $L_1 \ldots L_k$. Each variable L_i takes values in the finite space $\mathcal{L} \times \mathcal{T}$, where $\mathcal{L}$ is the set of entity identifiers and $\mathcal{T}$ is the set of all possible link types. Conceptually, $\mathcal{L} \times \mathcal{T}$ is just another vocabulary, and we can directly apply the cross-language techniques from section 5.3 to learn associations between the attributes of persons and their inter-relationships.

Semi-structured data

The idea of learning associations between entity attributes and the relational structure is certainly not novel. Researchers in the data mining community have proposed dozens of very effective and efficient algorithms for relational learning. We have no a-priori reason to believe that our model will outperform any of the existing solutions. However, our model does have one very important advantage over the classical mining algorithms. Most of these algorithms rely on *fully-structured* data representations. In order to learn that a certain person A is likely to have a *shared_office* link with person B we need to have access to entity attributes, such as age, academic position, research interests, etc. Classical mining algorithms will not work if instead of these attributes we provide them with the personal webpages of A and B, which are likely to be *unstructured* narratives. Our generative model has an advantage because it was specifically designed for learning associations from unstructured data. In fact, as far as our model is concerned, every aspect of the data is unstructured, some aspects just happen to have a very constrained vocabulary. Learning the presence of a *shared_office* link is no different than learning the presence of some Chinese word: both can be conditioned on a free-form English narrative. In a way, our model presents an interesting alternative to the popular mining approach: instead of *information extraction* followed by *data mining* we would directly learn associations over a mix of structured and unstructured components.

6.2.3 Order-dependent data

We believe it is possible to extend the generative model described in chapter 4 to order-dependent data. Consider a simple case of trying to learn a first-order dependence model, similar to a bi-gram. We have a training collection of observed bigrams $\mathcal{C}_{train} = \{(a_i, b_i) : i{=}1\ldots N\}$. The goal is to estimate the conditional probability distribution $P(a|b)$, representing the likelihood of observing a given that the previous word was b. A popular approach to estimating this distribution involves taking the maximum-likelihood estimate $\frac{\#(a,b)}{\#(b)}$, backing off to the unigram model and further smoothing the unigram counts. We suspect that a more powerful estimate may be constructed by setting $P(a|b) = \frac{P(a,b)}{\sum_a P(a,b)}$, where the distribution $P(a,b)$ is interpreted as a joint probability of seeing an Arabic word a together with a Bulgarian word b. The joint probability would be estimated as described in section 5.3 with $\mathcal{C}_{train}$ playing the role of a parallel corpus. The main advantage of the new estimate would be its ability to handle possible heterogeneity of word sequences.

6.2.4 Dirichlet kernels

In section 4.4 we described two types of kernel for our allocation model: the Dirac delta kernel and the Dirichlet kernel. Considerations of computational efficiency forced us to do most of our experiments with the simpler delta kernels. However, results reported in section 4.5 show that Dirichlet kernels represent a much more accurate instrument for modeling textual data. We have a strong interest in exploring the potential benefits of Dirichlet kernels in real retrieval scenarios. Explorations of this kind will require us to come up with a new set of algorithmic optimizations to bring the efficiency of our Dirichlet implementation a bit closer to the efficiency of a delta-based implementation.

References

1. J. Allan. *Introduction to Topic Detection and Tracking*, pages 1–16. Kluwer Academic Publishers, Massachusetts, 2002.
2. J. Allan, A. Bolivar, M. Connell, S. Cronen-Townsend, A. Feng, F. Feng, G. Kumaran, L. Larkey, V. Lavrenko, and H. Raghavan. UMass TDT 2003 research summary. In *NIST Topic Detection and Tracking Workshop*, Gaithersburg, MD, November 2003.
3. J. Allan, M. Connell, W.B. Croft, F.F. Feng, D. Fisher, and X. Li. INQUERY and TREC-9. In *Proceedings of the Ninth Text REtrieval Conference (TREC-9)*, pages 551–562, 2000.
4. J. Allan, R. Gupta, and V. Khandelval. Temporal summaries of news topics. In *Proceedings of the 24th International ACM SIGIR Conference on Research and Development in Information Retrieval*, pages 10–18, New Orleans, LA, September 2001.
5. J. Allan, J. Carbonell, G. Doddington, J. Yamron, and Y. Yang. Topic detection and tracking pilot study: Final report. In *Proceedings of DARPA Broadcast News Transcription and Understanding Workshop*, pages 194–218, 1998.
6. J. Allan, H. Jin, M. Rajman, C. Wayne, D. Gildea, V. Lavrenko, R. Hoberman, and D Caputo. Topic-based novelty detection. Clsp workshop final report, Johns Hopkins University, 1999.
7. J. Allan, V. Lavrenko, and H. Jin. First story detection in TDT is hard. In *Proceedings of the Ninth International Conference on Information and Knowledge Management CIKM*, pages 374–381, Wasnihgton, DC, 2000.
8. K. Barnard, P. Duygulu, N. Freitas, D. Forsyth, D. Blei, and M. Jordan. Matching words and pictures. *Journal of Machine Learning Research*, 3:1107–1135, 2003.
9. N. J. Belkin, R. N. Oddy, and H. M. Brooks. ASK for information retrieval: Part i. background and theory. *Journal of Documentation*, 38(2):61–71, 1982.
10. N. J. Belkin, R. N. Oddy, and H. M. Brooks. ASK for information retrieval: Part ii. results of a design study. *Journal of Documentation*, 38(3):145–164, 1982.
11. A. Berger and J. Lafferty. Information retrieval as statistical translation. In Hearst et al. [54], pages 222–229.
12. D.M. Blei, A.Y. Ng, and M.I. Jordan. Latent dirichlet allocation. In *Technical Report UCB//CSD-02-1194*, August 2002.

13. D. M. Blei, A. Y. Ng, and M. I. Jordan. Latent dirichlet allocation. *Journal of Machine Learning Research*, 3:993–1022, 2003.
14. D. Bodoff and S. E. Robertson. A new unified probabilistic model. *Journal of the American Society for Information Science*, 55(6):471–487, 2003.
15. A. Bookstein and D. Swanson. Probabilistic models for automatic indexing. *Journal of the American Society for Information Science*, 25(5):312–319, 1974.
16. A. Bookstein and D. Swanson. A decision theoretic foundation for indexing. *Journal of the American Society for Information Science*, 26:45–50, 1975.
17. P. F. Brown, J. Cocke, S. A. Della Pietra, V. J. Della Pietra, F. Jelinek, J. D. Lafferty, R. L. Mercer, and P. S. Roossin. A statistical approach to machine translation. *Computational Linguistics*, 16(2):79–85, 1990.
18. P. D. Bruza. *Stratified information disclosure: A synthesis between hypermedia and information retrieval.* PhD dissertation, University of Nijmegen, Nijmegen, The Netherlands, 1993.
19. P. D. Bruza and T. P. van der Weide. The modeling and retrieval of documents using index compression. *SIGIR Forum*, 25(2):91–103, 1991.
20. P. D. Bruza and T. P. van der Weide. Stratified hypermedia structures for information disclosure. *The Computer Journal*, 35(3):208–220, 1992.
21. J. Carbonell and J. Goldstein. The use of MMR, diversity-based reranking for reordering documents and producing summaries. In *Proceedings of the International ACM SIGIR Conference on Research and Development in Information Retrieval*, 1998.
22. C. Carson, M. Thomas, S. Belongie, J.M. Hellerstein, and J. Malik. Blobworld: A system for region-based image indexing and retrieval. In *Proceedings of the Third International Conference on Visual Information Systems*, pages 509–516, 1999.
23. Stanley F. Chen and Joshua T. Goodman. An empirical study of smoothing techniques for language modeling. In *Proceedings of the 34th Annual Meeting of the ACL*, 1996.
24. C. Cieri, S. Strassel, D. Graff, N. Martey, K. Rennert, and M. Liberman. *Corpora for Topic Detection and Tracking*, pages 33–66. Kluwer Academic Publishers, Massachusetts, 2002.
25. W. Cooper. A definition of relevance for information retrieval. *Information Storage and Retrieval*, 7(1):19–37, 1971.
26. W. Cooper. Exploiting the maximum entropy principle to increase retrieval effectiveness. *Journal of the American Society for Information Science*, 34(2):31–39, 1983.
27. W. Cooper and P. Huizinga. The maximum entropy principle and its application to the design of probabilistic retrieval systems. *Information Technology Research and Development*, 1(2):99–112, 1982.
28. W. Cooper and M. Maron. Foundation of probabilistic and utility-theoretic indexing. *Journal of the Association for Computing Machinery*, 25(1):67–80, 1978.
29. W. S. Cooper. Some inconsistencies and misidentified modeling assumptions in probabilistic information retrieval. *ACM Transactions on Information Systems*, 13(1):100–111, January 1995.
30. E. Cosijn and P. Ingwersen. Dimensions of relevance. *Information Processing and Management*, 36:533–550, 2000.
31. F. Crestani and C. J. Van Rijsbergen. Information retrieval by logical imaging. *Journal of Documentation*, 51(1):3–17, 1995.

32. F. Crestani and C. J. Van Rijsbergen. Probability kinematics in information retrieval. In *Proceedings of the 18th International ACM SIGIR Conference on Research and Development in Information Retrieval*, pages 291–299, Seattle, WA, 1995.
33. W. B. Croft, editor. *Advances in Information Retrieval — Recent Research from the Center for Intelligent Information Retrieval*, volume 7 of *The Kluwer international series on information retrieval*. Kluwer Academic Publishers, Boston, MA, April 2000.
34. W. B. Croft and D. Harper. Using probabilistic models of document retrieval without relevance information. *Journal of Documentation*, 35:285–295, 1979.
35. W. B. Croft, D. J. Harper, D. H. Kraft, and J. Zobel, editors. *Proceedings of the Twenty-Fourth Annual International ACM-SIGIR Conference on Research and Development in Information Retrieval*, New Orleans, LA, September 2001. ACM Press.
36. W. B. Croft and C. J. Van Rijsbergen. An evaluation of Goffman's indirect retrieval method. *Information Processing ad Management*, 12(5):327–331, 1976.
37. S. Cronen-Townsend, Y. Zhou, and W. B. Croft. Predicting query performance. In *In the proceedings of the 25th International ACM SIGIR Conference on Research and Development in Information Retrieval*, pages 299–306, Tampere, Finland, August 2002.
38. C. Cuadra and R. Katter. Opening the black box of "relevance". *Journal of Documentation*, 23(4):291–303, 1967.
39. C. Cuadra and R. Katter. The relevance of relevance assessment. *Proceedings of the American Documentation Institute*, 4:95–99, 1967.
40. S. Dumais. Latent semantic indexing (lsi): Trec-3 report. In *TREC-3 Proceedings*, pages 219–230, Gaithersburg, Maryland, November 1994.
41. P. Duygulu, K. Barnard, N. Freitas, and D. Forsyth. Object recognition as machine translation: Learning a lexicon for a fixed image vocabulary. In *Proceedings of the Seventh European Conference on Computer Vision*, pages 97–112, 2002.
42. M. Eisenberg. *Magnitude estimation and the measurement of relevance*. PhD dissertation, Syracuse University, Syracuse, NY, 1986.
43. R. A. Fairthorne. *Implications of test procedures*, pages 109–113. Case Western Reserve University Press, Cleveland, 1963.
44. S. Feng, R. Manmatha, and V. Lavrenko. Multiple bernoulli relevance models for image and video annotation. In *Proceedings of the International Conference on Pattern Recognition (CVPR)*, Washington, DC, July 2004.
45. Jonathan G. Fiscus and George R. Doddington. *Topic Detection and Tracking Evaluation Overview*, pages 17–31. Kluwer Academic Publishers, Massachusetts, 2002.
46. D. J. Foskett. Classification and indexing in the social sciences. *Proceedings of ASLIB*, 22:90–100, 1970.
47. D. J. Foskett. A note on the concept of relevance. *Information Storage and Retrieval*, 8(2):77–78, 1972.
48. W. Goffman. An indirect method of information retrieval. *Information Storage and Retrieval*, 4:361–373, 1969.
49. D. Harman. Overview of the TREC 2002 novelty track. In *Proceedings of TREC 2002 (notebook)*, pages 17–28, 2002.
50. D.J. Harper. *Relevance feedback in document retrieval systems*. PhD thesis, University of Cambridge, UK, February 1980.

51. D. J. Harper and C. J. van Rijsbergen. An evaluation of feedback in document retrieval using co-occurrence data. *Journal of Documentation*, 34:189–216, 1978.
52. S. P. Harter. A probabilistic approach to automatic keyword indexing. *Journal of the American Society for Information Science*, 26(4 and 5):Part I: 197–206; Part II: 280–289, 1975.
53. S. P. Harter Variations in relevance assessments and the measurement of retrieval effectiveness. *Journal of the American Society for Information Science*, 47:37–49, 1996.
54. M. Hearst, F. Gey, and R. Tong, editors. *Proceedings of the Twenty-Second Annual International ACM-SIGIR Conference on Research and Development in Information Retrieval*, Berkeley, CA, August 1999. ACM Press.
55. M. Hearst and J. Pedersen. Re-examining the cluster hypothesis: scatter/gather on retrieval results. In *Proceedings of the International ACM SIGIR Conference on Research and Development in Information Retrieval*, pages 76–84, 1996.
56. D. Hiemstra. *Using Language Models for Information Retrieval.* PhD dissertation, University of Twente, Enschede, The Netherlands, January 2001.
57. D. Hiemstra and F. de Jong Disambiguation strategies for cross-language information retrieval. In S. Abiteboul and A.-M. Vercoustre, editors, *Proceedings of the Third European Conference on Research and Advanced Technology for Digital Libaries, ECDL'99*, volume 1696 of *Lecture Notes in Computer Science*, pages 274–293. Springer-Verlag, Paris, September 1999.
58. T. Hoffmann. Probabilistic latent semantic indexing. In *Proceedings on the 22nd annual international ACM SIGIR conference*, pages 50–57, 1999.
59. P. Ingwersen. *Information Retrieval Interaction.* Taylor Graham, London, 1992.
60. J. Janes. The binary nature of continuous relevance judgments: A case study of users' perceptions. *Journal of the American Society for Information Science*, 42(10):754–756, 1991.
61. J. Janes. On the distribution of relevance judgments. In *Proceedings of the American Society for Information Science*, pages 104–114, Medford, NJ, 1993.
62. N. Jardine and C. J. Van Rijsbergen. The use of hierarchical clustering in information retrieval. *Information Storage and Retrieval*, 7:217–240, 1971.
63. J. Jeon, V. Lavrenko, and R. Manmatha. Automatic image annotation and retrieval using cross-media relevance models. In *In the proceedings of the 26th International ACM SIGIR Conference on Research and Development in Information Retrieval*, pages 119–126, Toronto, Canada, August 2003.
64. P. Kantor. Maximum entropy and the optimal design of automated information retrieval systems. *Information Technology Research and Development*, 3(2):88–94, 1984.
65. R. Katter. The influence of scale form on relevance judgment. *Information Storage and Retrieval*, 4(1):1–11, 1968.
66. S. Kullback, M. Kupperman, and H. H. Ku. Tests for contingency tables and markov chains. *Technometrics*, 4(4):573–608, 1962.
67. O. Kurland and L. Lee. Corpus structure, language models, and ad hoc information retrieval. In *In the proceedings of the 27th International ACM SIGIR Conference on Research and Development in Information Retrieval*, pages 194–201, Sheffield, UK, July 2004.

68. J. Lafferty and C. Zhai. Document language models, query models, and risk minimization for information retrieval. In Croft et al. [35], pages 111–119.
69. J. Lafferty and ChengXiang Zhai. *Probabilistic relevance models based on document and query generation*, pages 1–10. Kluwer Academic Publishers, Dordrecht, The Netherlands, 2003.
70. M. Lalmas and C. J. Van Rijsbergen. A logical model of information retrieval based on situation theory. In *Proceedings of the BCS 14th Information Retrieval Colloquium*, pages 1–13, Lancaster, UK, 1992.
71. M. Lalmas and C. J. Van Rijsbergen. Situation theory and Dempster-Shafer's theory of evidence for information retrieval. In *Proceedings of Workshop on Incompleteness and Uncertainty in Information Systems*, pages 62–67, Concordia University, Montreal, Canada, 1993.
72. M. Lalmas and C. J. Van Rijsbergen. Information calculus for information retrieval. *Journal of the American Society for Information Science*, 47:385–398, 1996.
73. F. W. Lancaster. *Information retrieval systems: Characteristics, testing and evaluation*. John Wiley and Sons, New York, 1979.
74. V. Lavrenko, J. Allan, E. DeGuzman, D. LaFlamme, V. Pollard, and S. Thomas. Relevance models for topic detection and tracking. In *Proceedings of Human Language Technologies Conference, HLT 2002*, pages 104–110, 2002.
75. V. Lavrenko and W. B. Croft. *Relevance Models in Information Retrieval*, pages 11–56. Kluwer Academic Publishers, Dordrecht, The Netherlands, 2003.
76. V. Lavrenko, S. Feng, and R. Manmatha. Statistical models for automatic video annotation and retrieval. In *Proceedings of the International Conference on Acoustics, Speech and Signal Processing (ICASSP)*, Montreal, Canada, May 2004.
77. V. Lavrenko, R. Manmatha, and J. Jeon. A model for learning the semantics of pictures. In *Proceedings of the Seventeenth Annual Conference on Neural Information Processing Systems (NIPS)*, Vancouver, Canada, December 2003.
78. V. Lavrenko, T. Rath, and R. Manmatha. Holistic word recognition for handwritten historical documents. In *Preceedings of the International Workshop on Document Image Analysis for Libraries (DIAL)*, Palo Alto, CA, January 2004.
79. V. Lavrenko, X. Yi, and J. Allan. Information retrieval on empty fields. In *Proceedings of Human Language Technologies Conference (NAACL-HLT)*, pages 89–96, Rochester, NY, April 2007.
80. J. Lee and P. Kantor. A study of probabilistic information retrieval systems in the case of inconsistent expert judgment. *Journal of the American Society for Information Science*, 42(3):166–172, 1991.
81. D. Lenat. CYC: A large scale investment in knowledge infrastructure. *Communications of the ACM*, 38(11):33–38, 1995.
82. A. Leuski. Evaluating document clustering for interactive information retrieval. In *Proceedings of CIKM 2001 conference*, pages 33–40, 2001.
83. Tipster Volume 1. CDROM available from Linguistic Data Consortium, University of Pennsylvania, 1993, Revised March 1994. http://morph.ldc.upenn.edu/Catalog/LDC93T3B.html.
84. R. Manmatha and W. B. Croft. *Word spotting: Indexing handwritten manuscripts*, pages 43–64. AAAI/MIT Press, Dordrecht, The Netherlands, 1997.

85. M. E. Maron and J. L. Kuhns. On relevance, probabilistic indexing and information retrieval. *Journal of the Association for Computing Machinery*, 7(3):216–244, 1960.
86. A. Martin, G. Doddington, T. Kamm, and M. Ordowski. The DET curve in assessment of detection task performance. In *EuroSpeech*, pages 1895–1898, 1997.
87. C. Meghini, F. Sebastiani, U. Straccia, and C. Thanos. A model of information retrieval based on terminological logic. In *Proceedings of the 16th International ACM SIGIR Conference on Research and Development in Information Retrieval*, pages 298–307, Pittsburgh, PA, 1993.
88. D. Metzler, V. Lavrenko, and W. B. Croft. Formal multiple-Bernoulli models for language modeling. In *Proceedings of the 27th International ACM SIGIR Conference on Research and Development in Information Retrieval*, Sheffield, UK, July 2004.
89. D. Miller, T. Leek, and R. Schwartz. BBN at TREC7: Using hidden markov models for information retrieval. In Voorhees and Harman [145], pages 133–142.
90. D. R. H. Miller, T. Leek, and R.M. Schwartz. A hidden markov model information retrieval system. In Hearst et al. [54], pages 214–221.
91. S. Mizarro Relevance: The whole history. *Journal of the American Society for Information Science*, 48(9):810–832, 1997.
92. S. Mizarro How many relevances in information retrieval? *Interacting with Computers*, 10(3):305–322, 1998.
93. Y. Mori, H. Takahashi, and R. Oka. Image-to-word transformation based on dividing and vector quantizing images with words. In *Proceedings of the First International Workshop on Multimedia Intelligent Storage and Retrieval Management MISRM'99*, 1999.
94. R. Nallapati and J. Allan. Capturing term dependencies using a sentence tree based language model. In *Proceedings of CIKM 2002 conference*, pages 383–390, 2002.
95. R. Nallapati and J. Allan. An adaptive local dependency language model: Relaxing the naive Bayes assumption. In *SIGIR'03 Workshop on Mathematical Methods in Information Retrieval*, Toronto, Canada, 2003.
96. R. Nallapati, W. B. Croft, and J. Allan. Relevant query feedback in statistical language modeling. In *Proceedings of CIKM 2003 conference*, pages 560–563, 2003.
97. J. Neyman and E. S. Pearson. On the use and interpretation of certain test criteria for purposes of statistical inference: Part ii. *Biometrika*, 20A(3/4):263–294, 1928.
98. J. Y. Nie. An outline of a general model for information retrieval. In *Proceedings of the 11th International ACM SIGIR Conference on Research and Development in Information Retrieval*, pages 495–506, Grenoble, France, 1988.
99. J. Y. Nie An information retrieval model based on modal logic. *Information Processing and Management*, 25(5):477–491, 1989.
100. J.Y. Nie. Towards a probabilistic modal logic for semantic-based information retrieval. In *Proceedings of the 15th International ACM SIGIR Conference on Research and Development in Information Retrieval*, pages 140–151, Copenhagen, Denmark, 1992.
101. J. Y. Nie, M. Brisebois, and F. Lepage Information retrieval is counterfactual. *The Computer Journal*, 38(8):643–657, 1995.

102. K. Nigam, A. McCallum, S. Thrun, and T. Mitchell Text classification from labeled and unlabeled documents using em. *Machine Learning Journal*, 39(2/3):103–134, 2000.
103. NIST. Proceedings of the tdt 2001 workshop. Notebook publication for participants only, November 2001.
104. R. Papka. *On-line New Event Detection, Clustering, and Tracking*. PhD dissertation, TR99-45, University of Massachusetts, Amherst, MA, 1999.
105. J. M. Ponte. *A language modeling approach to information retrieval.* Phd dissertation, University of Massachusets, Amherst, MA, September 1998.
106. J. M. Ponte and W. B. Croft. A language modeling approach to information retrieval. In W. B. Croft, A. Moffat, C. J. van Rijsbergen, R. Wilkinson, and J. Zobel, editors, *Proceedings of the Twenty-First Annual International ACM-SIGIR Conference on Research and Development in Information Retrieval*, pages 275–281, Melbourne, Australia, August 1998. ACM Press.
107. T. Rath, V. Lavrenko, and R. Manmatha. A statistical approach to retrieving historical manuscript images. In *CIIR Technical Report MM-42*, 2003.
108. S. Robertson. On Bayesian models and event spaces in information retrieval. Presented at SIGIR 2002 Workshop on Mathematical/Formal Models in Information Retrieval, 2002.
109. S. Robertson. The unified model revisited. Presented at SIGIR 2003 Workshop on Mathematical/Formal Models in Information Retrieval, 2003.
110. S. Robertson and J. Bovey. Statistical problems in the application of probabilistic models to information retrieval. Technical Report 5739, British Library Research and Development Department, 1991.
111. S. Robertson and D. Hiemstra. Language models and probability of relevance. In *Proceedings of the Workshop on Language Modeling and Information Retrieval*, pages 21–25, Pittsburgh, PA, May 2001.
112. S. Robertson and S. Walker. On relevance weights with little relevance information. In *Proceedings of the 20th International ACM SIGIR Conference on Research and Development in Information Retrieval*, pages 16–24, 1997.
113. S. E. Robertson. The probabilistic character of relevance. *Information Processing and Management*, 13:247–251, 1977.
114. S. E. Robertson. The probability ranking principle in IR. *Journal of Documentation*, 33:294–304, 1977. Reprinted in [133].
115. S. E. Robertson, M. E. Maron, and W. S. Cooper. Probability of relevance: a unification of two competing models for document retrieval. *Information Technology: Research and Development*, 1:1–21, 1982.
116. S. E. Robertson, M. E. Maron, and W. S. Cooper. *The unified probabilistic model for IR*, pages 108–117. Springer-Verlag, Berlin, Germany, 1983.
117. S. E. Robertson and K. Sparck Jones. Relevance weighting of search terms. *Journal of the American Society for Information Science*, 27:129–146, 1976. Reprinted in [149].
118. S. E. Robertson and S. Walker. Some simple effective approximations to the 2–poisson model for probabilistic weighted retrieval. In W. B. Croft and C. J. van Rijsbergen, editors, *Proceedings of the Seventeenth Annual International ACM-SIGIR Conference on Research and Development in Information Retrieval*, pages 232–241, Dublin, Ireland, July 1994. Springer-Verlag.
119. S. E. Robertson, S. Walker, and M. M. Beaulieu. Okapi at TREC-7: automatic ad hoc, filtering, VLC and interactive track. In Voorhees and Harman [145], pages 253–264.

120. T. Saracevic Relevance: A review of and a framework for the thinking on the notion in information science. *Journal of the American Society for Information Science*, 26(6):321–343, 1975.
121. T. Saracevic. Relevance reconsidered'96. In *Information Science: Integration in Perspective*, pages 201–218, Royal School of Library and Information Science, Copenhagen, Denmark, 1996.
122. F. Sebastiani. A probabilistic terminological logic for information retrieval. In *Proceedings of the 17th International ACM SIGIR Conference on Research and Development in Information Retrieval*, pages 122–130, Dublin, Ireland, 1994.
123. F. Sebastiani. On the role of logic in information retrieval. *Information Processing and Management*, 38(1):1–18, 1998.
124. J. Shi and J. Malik. Normalized cuts and image segmentation. *IEEE Transactions on Pattern Analysis and Machine Intelligence*, 22(8):888–905, 2000.
125. B. Silverman. *Density Estimation for Statistics and Data Analysis*, pages 75–94. CRC Press, 1986.
126. F. Song and W.B. Croft. A general language model for information retrieval. In *Proceedings of Eighth International Conference on Information and Knowledge Management, CIKM'99*, 1999.
127. F. Song and W.B. Croft. A general language model for information retrieval. In *Proceedings of the Twenty-Second Annual International ACM-SIGIR Conference on Research and Development in Information Retrieval, SIGIR'99*, pages 279–280, 1999.
128. K. Sparck Jones. A statistical interpretation of term specificity and its application in retrieval. *Journal of Documentation*, 28:11–21, 1972. Reprinted in [149].
129. K. Sparck Jones. LM vs. PM: Where's the relevance? In *Proceedings of the Workshop on Language Modeling and Information Retrieval*, pages 12–15, Pittsburgh, PA, May 2001.
130. K. Sparck Jones, S. Robertson, D. Hiemstra, and H. Zaragoza. *Language Modeling and Relevance*, pages 57–72. Kluwer Academic Publishers, Dordrecht, The Netherlands, 2003.
131. K. Sparck Jones, S. Walker, and S. Robertson. A probabilistic model of information retrieval: development and comparative experiments. part 1. *Information Processing and Management*, 36:779–808, 2000.
132. K. Sparck Jones, S. Walker, and S. Robertson. A probabilistic model of information retrieval: development and comparative experiments. part 2. *Information Processing and Management*, 36:809–840, 2000.
133. K. Sparck Jones and P. Willett, editors. *Readings in information retrieval.* Multimedia Information and Systems. Morgan Kaufmann, San Francisco, CA, 1997.
134. H.R. Turtle. *Inference Network for Document Retrieval.* PhD dissertation, University of Massachusets, Amherst, MA, 1990.
135. H. R. Turtle and W. B. Croft. Evaluation of an inference network-based retrieval model. *ACM Transactions on Information Systems*, 9(3):187–222, July 1991.
136. C. J. van Rijsbergen. A theoretical basis for the use of co-occurrence data in information retrieval. *Journal of Documentation*, 33:106–119, 1977.
137. C. J. van Rijsbergen. *Information retrieval.* Butterworth & Co (Publishers) Ltd, London, UK, second edition, 1979.

138. C.J. Van Rijsbergen. A new theoretical framework for information retrieval. In *Proceedings of the International ACM SIGIR Conference on Research and Development in Information Retrieval*, page 200, Pisa, Italy, 1986.
139. C. J. van Rijsbergen. A non-classical logic for information retrieval. *The Computer Journal*, 29:481–485, 1986.
140. C. J. Van Rijsbergen. Towards an information logic. In *Proceedings of the 12th International ACM SIGIR Conference on Research and Development in Information Retrieval*, pages 77–86, Cambridge, MA, 1989.
141. C. J. Van Rijsbergen and W. B. Croft. Document clustering: An evaluation of some experiments with the Cranfield 1400 collection. *Information Processing and Management*, 11:171–182, 1975.
142. B. C. Vickery. The structure of information retrieval systems. *Proceedings of the International Conference on Scientific Information*, 2:1275–1290, 1959.
143. B. C. Vickery. Subject analysis for information retrieval. *Proceedings of the International Conference on Scientific Information*, 2:855–865, 1959.
144. E. Voorhees. *The effectiveness and efficiency of agglomerative hierarchic clustering in document retrieval.* PhD dissertation, TR85-705, Cornell University, NY, 1986.
145. E.M. Voorhees and D.K. Harman, editors. *Proceedings of the Seventh Text REtrieval Conference (TREC-7)*, Gaithersburg, MD, November 1998. National Institute of Standards and Technology (NIST) and Defense Advanced Research Projects Agency (DARPA), Department of Commerce, National Institute of Standards and Technology.
146. E.M. Voorhees and D.K. Harman, editors. *Proceedings of the Ninth Text REtrieval Conference (TREC-9)*, Gaithersburg, MD, November 2000. Department of Commerce, National Institute of Standards and Technology.
147. Ronald E. Walpole and Raymond H. Myers. *Probability and Statistics for Engineers and Scientists*, pages 108–109. MacMillan Publishing Company, New York, 1989.
148. S.S. Wilks. The likelihood test of independence in contingency tables. *The Annals of Mathematical Statistics*, 6(4):190–196, 1935.
149. P. Willett, editor. *Document Retrieval Systems*, volume 3 of *Foundations of Information Science.* Taylor Graham, London, UK, 1988.
150. P. Wilson. Situational relevance. *Information Storage and Retrieval*, 9(8):457–471, 1973.
151. J. Xu and W. B. Croft. Query expansion using local and global document analysis. In H.-P. Frei, D. Harman, P. Schäuble, and R. Wilkinson, editors, *Proceedings of the Nineteenth Annual International ACM-SIGIR Conference on Research and Development in Information Retrieval*, pages 4–11, Zurich, Switzerland, August 1996. ACM Press.
152. J. Xu and W. B. Croft. Cluster-based language models for distributed retrieval. In *Proceedings of the 22nd International ACM SIGIR Conference on Research and Development in Information Retrieval (SIGIR 99)*, pages 15–19, Berkeley, CA, August 1999.
153. J. Xu and R. Weischedel. TREC-9 cross-lingual retrieval at BBN. In Voorhees and Harman [146], pages 106–116.
154. J. Xu, R. Weischedel, and C. Nguyen. Evaluating a probabilistic model for cross-lingual information retrieval. In Croft et al. [35], pages 105–110.

155. X. Yi, J. Allan, and V. Lavrenko. Discovering missing values in semi-structured databases. In *Proceedings of RIAO 2007 - 8th Conference - Large-Scale Semantic Access to Content (Text, Image, Video and Sound)*, page Paper No. 113, Pittsburgh, PA, May 2007.
156. C. Yu, C. Buckley, and G. Salton. A generalized term dependence model in information retrieval. *Information Technology Research and Development*, 2:129–154, 1983.
157. C. Zhai. *Risk Minimization and Language Modeling in Text Retrieval.* PhD dissertation, Carnegie Mellon University, Pittsburgh, PA, July 2002.
158. C. Zhai and J. Lafferty. A study of smoothing methods for language models applied to ad hoc information retrieval. In Croft et al. [35], pages 334–342.
159. C. Zhai and J. Lafferty. Two-stage language models for information retrieval. In *Proceedings of the Twenty-Fifth Annual International ACM-SIGIR Conference on Research and Development in Information Retrieval*, pages 49–56, Tampere, Finland, 2002.

Index